SUSTAINABLE FOOD
AND BEVERAGE INDUSTRIES

Assessments and Methodologies

SUSTAINABLE FOOD AND BEVERAGE INDUSTRIES

Assessments and Methodologies

Edited by
Gabriela Ionescu, PhD

Apple Academic Press Inc. | Apple Academic Press Inc.
3333 Mistwell Crescent | 9 Spinnaker Way
Oakville, ON L6L 0A2 | Waretown, NJ 08758
Canada | USA

©2016 by Apple Academic Press, Inc.

First issued in paperback 2021

Exclusive worldwide distribution by CRC Press, a member of Taylor & Francis Group

No claim to original U.S. Government works

ISBN 13: 978-1-77463-701-2 (pbk)
ISBN 13: 978-1-77188-410-5 (hbk)

Library and Archives Canada Cataloguing in Publication

Sustainable food and beverage industries : assessments and methodologies/edited by Gabriela Ionescu, PhD.

Includes bibliographical references and index.
Issued in print and electronic formats.
ISBN 978-1-77188-410-5 (hardcover).--ISBN 978-1-77188-411-2 (pdf)
1. Food industry and trade--Environmental aspects--Case studies. 2. Food supply--Environmental aspects--Case studies. 3. Food processing plants--Environmental aspects--Case studies. 4. Beverage industry--Environmental aspects--Case studies. 5. Waste products--Case studies. 6. Packaging waste--Case studies. 7. Social responsibility of business--Case studies. I. Ionescu, Gabriela, author, editor

TD195.F57S88 2016   664.0028'6   C2016-900381-7   C2016-900382-5

Library of Congress Cataloging-in-Publication Data

Names: Ionescu, Gabriela, editor.
Title: Sustainable food and beverage industries : assessments and methodologies / editor, Gabriela Ionescu, PhD.
Description: 1st Edition. | New Jersey : Apple Academic Press, 2016. |
Includes bibliographical references and index.
Identifiers: LCCN 2016002071 (print) | LCCN 2016007322 (ebook) | ISBN 9781771884105 (hardcover : alk. paper) | ISBN 9781771884112 ()
Subjects: LCSH: Food industry and trade. | Beverage industry.
Classification: LCC HD9000.5 .S833 2016 (print) | LCC HD9000.5 (ebook) | DDC 338.4/7664--dc23
LC record available at http://lccn.loc.gov/2016002071

Apple Academic Press also publishes its books in a variety of electronic formats. Some content that appears in print may not be available in electronic format. For information about Apple Academic Press products, visit our website at **www.appleacademicpress.com** and the CRC Press website at **www.crcpress.com**

About the Editor

GABRIELA IONESCU, PhD

Dr. Gabriela Ionescu obtained her PhD in Power Engineering from Politehnica University of Bucharest and in Environmental Engineering from University of Trento. She is currently a member of the Department of Energy Production and Use at the Politehnica University of Bucharest and collaborator of the Department of Civil, Environmental and Mechanical Engineering at the University of Trento. She has done prolific research and has been published multiple times in areas related to energy efficiency, waste and wastewater management, energy conservation, life-cycle assessment, environmental analysis, and sustainability feasibility studies.

Contents

Acknowledgment and How to Cite

The editor and publisher thank each of the authors who contributed to this book. The chapters in this book were previously published elsewhere. To cite the work contained in this book and to view the individual permissions, please refer to the citation at the beginning of each chapter. Each chapter was carefully selected by the editor; the result is a book that looks at the sustainable food and beverage industry from a variety of perspectives. The chapters included are broken into seven sections, which describe the following topics:

Part I: Overview of Food Production and Supply Chains
- Chapter 1 summarizes key highlights from the book entitled Green Technologies in Food Production and Processing, which include key drivers of the evolution in the food supply chain and in-depth description of food production and processing using the life cycle assessment (LCA) tool; approaches to improve food production practices; sustainable food processing approaches; emerging analytical techniques for sustainable research and development; challenges associated with the use of agricultural resources to grow biofuels and bio-based products; technologies to reduce the generation of process-induced toxins; social factors that influence consumer perceptions about some of the current and emerging agri-food technologies; and the need and importance of biodiversity in maintaining sustainable human diets. This forms a good foundation for the chapters that follow.
- The authors of chapter 2 (Iakova et al.) create a framework intended to optimize the agrifood supply chain design, planning, and operations through the implementation of appropriate green supply chain management and logistics principles, while minimizing the environmental burden.

Part II: Dairy Industry
- In chapter 3, O'Brien and his colleagues examine the factors that cause variation among dairy farms' carbon footprints.
- The study in chapter 4 demonstrates the fermentation of non-whey, lactose-rich industrial diary waste for acetone, butanol, and ethanol production.

- The authors of chapter 5 (Kim et al.) use life cycle assessments to create recommendations to help the cheese industry engage in sustainable practices and reduce environmental impacts.

Part III: Meat Industry
- In chapter 6, Dyer and his colleagues quantify the green house gas emissions from the Canadian sheep industry versus the beef industry. Although their research focuses on Canada, they present valuable implications for other meat industries throughout the world, as well as questions that call for further research.

Part IV: Coffee and Tea Industries
- Chapter 7 focuses on sustainability in the American fair-trade coffee industry. The authors' findings have limited applicability to other sectors beyond coffee, but Howard and Jaffee do suggest hypotheses that might be tested in other industries.
- Chapter 8 presents a case study that suggests ways to promote the productivity and sustainability of tea plantation soils at the farm scale through an integrated natural resource management system, while using fertilizers from organic sources.

Part V: Food and Beverage Waste Products
- Chapter 9 presents the results of an experiment that examined the physical-chemical characterization of meat processing industry residues (bones), in order to assess their energetic potential as a renewable energy source. The authors also use their experimental analysis to identify possible energy conversion solutions.
- Chapter 10 contributes to the study of anaerobic sequencing batch reactor viability in the stillage treatment sector.

Part VI: Food Processing and Packaging
- Chapter 11 is a good summary of the integrated production and treatment biotech process, an engineering strategy for the sustainable production of renewable resources from waste organic materials.

Part VII: Concluding Implications
- In chapter 12, the authors (Reynolds et al.) review the current environmental impact assessment and life cycle analysis (LCA) literature around the environmental impacts of dietary recommendations, focusing on collating the environmental evidence behind three pieces of dietary advice that are debated in current environmental impact assessment and LCA literature.

List of Contributors

Ch. Achillas
School of Economics, Business Administration & Legal Studies, International Hellenic University, 14th km Thessaloniki-Moudania, 57001 Thermi, Greece

F. Anastasiadis
School of Economics, Business Administration & Legal Studies, International Hellenic University, 14th km Thessaloniki-Moudania, 57001 Thermi, Greece

Yves Arcand
Food Research and Development Centre, Agriculture and Agri-Food Canada

Ashok Kumar Bharathidasan
Department of Food, Agricultural and Biological Engineering, The Ohio State University (OSU) and Ohio Agricultural Research and Development Center (OARDC), Wooster, USA

John Boland
Centre for Industrial and Applied Mathematics, the Barbara Hardy Institute, University of South Australia, Mawson Lakes Boulevard, Mawson Lakes, SA 5095, Australia

Joyce I. Boye
Food Research and Development Centre, Agriculture and Agri-Food Canada

Padraig Brennan
Bord Bia, Clanwilliam Court, Lower Mount Street, Dublin 2, Ireland

Jonathan David Buckley
Nutritional Physiology Research Centre, Sansom Institute for Health Research, University of South Australia, Mawson Lakes Boulevard, Mawson Lakes, SA 5095, Australia

Simona Ciută
Asist., Power Engineering Faculty, University POLITEHNICA of Bucharest, Romania

Katrina Cornish
Department of Horticulture & Crop Sciences, The Ohio State University (OSU) and Ohio Agricultural Research and Development Center (OARDC), Wooster, USA

Raymond L. Desjardins
Agriculture and Agri-Food Canada, 960 Carling Avenue, Ottawa, Ontario, Canada

James A. Dyer
Agro-environmental Consultant, 122 Hexam Str., Cambridge, Ontario, Canada

Thaddeus Chukwuemeka Ezeji
Department of Animal Sciences, The Ohio State University (OSU) and Ohio Agricultural Research and Development Center (OARDC), Wooster, USA

Md. Abdul Halim
SRDI, District office, Dinajpur, Bangladesh

Philip H. Howard
Department of Community, Agriculture, Recreation and Resource Studies, Michigan State University, 480 Wilson RD, Rm. 316, East Lansing, MI 48824, USA

James Humphreys
Livestock Systems Research Department, Animal & Grassland Research and Innovation Centre, Teagasc, Moorepark, Fermoy, County Cork, Ireland

E. Iakovou
Laboratory of Logistics and Supply Chain Management, Department of Mechanical Engineering, Aristotle University of Thessaloniki, 54124 Thessaloniki, Greece

Gabriela Ionescu
Asist., Power Engineering Faculty, University POLITEHNICA of Bucharest, Romania

Daniel Jaffee
Department of Sociology, Washington State University, 14204 NE Salmon Creek Avenue, Vancouver, WA 98686, USA

Bo Jin
School of Chemical Engineering, The University of Adelaide, Adelaide, SA 5095, Australia

Md. Kamaruzzaman
SRDI, Regional office, Rajshahi

Daesoo Kim
Department of Chemical Engineering, University of Arkansas, Fayetteville, AR 72701, USA

Cosmin Mărculescu
Lecturer, Power Engineering Faculty, University POLITEHNICA of Bucharest, Romania

Franco Milani
Jeneil Biotech, Inc., 400 N. Dekora Woods Blvd., Saukville, WI 53080, USA

Greg Norris
Center for Health and the Global Environment, Harvard School of Public Health, Harvard University, Boston, MA 02115, USA

Darin Nutter
Department of Mechanical Engineering, University of Arkansas, Fayetteville, AR 72701, USA

Donal O'Brien
Livestock Systems Research Department, Animal & Grassland Research and Innovation Centre, Teagasc, Moorepark, Fermoy, County Cork, Ireland

Elena Cristina Rada
University of Trento, Department of Civil and Environmental Engineering, Via Mesiano 77, 38123 Trento, Italy.

Marco Ragazzi
University of Trento, Department of Civil and Environmental Engineering, Via Mesiano 77, 38123 Trento, Italy.

Christian John Reynolds
Centre for Industrial and Applied Mathematics, the Barbara Hardy Institute, University of South Australia, Mawson Lakes Boulevard, Mawson Lakes, SA 5095, Australia and Integrated Sustainability Analysis, University of Sydney, Sydney, 2000, Australia

Eimear Ruane
Livestock Systems Research Department, Animal & Grassland Research and Innovation Centre, Teagasc, Moorepark, Fermoy, County Cork, Ireland

Laurence Shalloo
Livestock Systems Research Department, Animal & Grassland Research and Innovation Centre, Teagasc, Moorepark, Fermoy, County Cork, Ireland

Md. Noor-E-Alam Siddique
Soil Resource Development Institute (SRDI), Ministry of Agriculture, District Office, Pabna

Constantin Stan
Asist., Power Engineering Faculty, University POLITEHNICA of Bucharest, Romania

Jakia Sultana
Organic Agriculture, Wageningen University, the Netherlands

Vincenzo Torretta
Insubria University, Department of Science and High Technology, Via Vico 46, 21100 Varese, Italy.

Greg Thoma
Department of Chemical Engineering, University of Arkansas, Fayetteville, AR 72701, USA

Rick Ulrich
Department of Chemical Engineering, University of Arkansas, Fayetteville, AR 72701, USA

Victor Ujor
Department of Animal Sciences, The Ohio State University (OSU) and Ohio Agricultural Research and Development Center (OARDC), Wooster, USA

Xavier P. C. Vergé
Consultant to AAFC, Ottawa, Ontario, Canada

D. Vlachos
Laboratory of Logistics and Supply Chain Management, Department of Mechanical Engineering, Aristotle University of Thessaloniki, 54124 Thessaloniki, Greece

Philip Weinstein
School of Pharmacy and Medical Sciences, Division of Health Sciences, University of South Australia, Mawson Lakes Boulevard, Mawson Lakes, SA 5095, Australia

Devon E. Worth
Agriculture and Agri-Food Canada, 960 Carling Avenue, Ottawa, Ontario, Canada

Introduction

Food and beverage industries are intrinsically and directly dependent on the Earth's food, water, and energy, making this industry sector particularly vulnerable to climate change and other environmental challenges. While population growth puts increased pressure on the system to deliver more food, the impacts of climate change on agriculture jeopardizes food production. At the same time, consumers are increasingly conscious of where their food comes from and how it is delivered. The demand for organic food, local foods, and other alternatives puts unique pressures on these industries. In order to appeal to a more health-conscious market, many companies are seeking to change the way they grow, process, and transport their products.

A recent study [1] reveals a definite sustainability trend in the food and beverage sector. The report examined twenty-four food and beverage companies, including the Campbell Soup Company, PepsiCo, Kraft Foods, and The Coca Cola Company. The companies' current sustainability programs were evaluated based on how prepared they were to meet the sustainability challenges of the twenty-first century. The results indicate that food and beverage industries are leading other sectors in some areas (engaging stakeholders in the sustainability movement and disclosure to investors); 79 percent are reducing their greenhouse gas (GHG) emissions; 60 percent are actively working to involve their employees in sustainability efforts; but only 25 percent of the reported companies are using renewable energy sources in their operations, and an even smaller percentage are working to reduce the environmental impact of their transportation networks.

The food and beverage sectors' ability to adapt to the demands of the twenty-first century, adopting effective sustainable practices, will depend on a solid research foundation. Researchers must develop processes that will efficiently and effectively allow industries to meet sustainability goals. Carbon footprints and GHG must be reduced at all levels of production, including agriculture, processing plants, and supply chains. Waste

products must be managed sustainably at all these leves as well, with minimum impact on the environment.

The articles in this compendium have been selected because they offer
an up-to-date broad foundation for the vital challenges facing the food and
beverage sector. This research—and the research that will build on it in the
years to come—offers these industries the direction they need to meet the
twenty-first century's urgent sustainability demands.

REFERENCES

1. Ceres, "Ceres Roadmap for Sustainability," 2014. http://www.ceres.org/roadmap-
    assessment/progress-report/progress-report. Accessed October 5, 2015.

Gabriela Ionescu

Finding a balance between food supply and demand in a manner that
is sustainable and which ensures the long-term survival of the human species will be one of the most important challenges for humankind in the
coming decades. Global population growth in the last several centuries
with the attendant demands resulting from industrialization has made the
need for food production and processing an important issue. This need is
expected to increase in the next half century when the population of the
world exceeds 9 billion. Environmental concerns related to food production and processing which require consideration include land use change
and tremendous reduction in biodiversity, aquatic eutrophication by nitrogenous and phosphorus substances caused by over-fertilization, climate
change, water shortages due to irrigation, ecotoxicity, and human effects
of pesticides, among others. Chapter 1, by Boye and Arcand, summarizes
key highlights from the recently published book entitled *Green Technologies in Food Production and Processing,* which provides a comprehensive
summary of the current status of the agriculture and agri-food sectors in
regard to environmental sustainability and material and energy steward-

ship and further provides strategies that can be used by industries to enhance the use of environmentally friendly technologies for food production and processing.

Agrifood sector is one of the most important economic and political areas within the European Union, with key implications for sustainability such as the fulfillment of human needs, the support of employment and economic growth, and its impact on the natural environment. Growing environmental, social and ethical concerns and increased awareness of the impacts of the agrifood sector have led to increased pressure by all involved supply chain stakeholders, while at the same time the European Union has undertaken a number of relevant regulatory interventions. Chapter 2, by Iakovou and colleagues, aims to present a methodological framework for the design of green supply chains for the agrifood sector. The framework aims towards the optimization of the agrifood supply chain design, planning and operations through the implementation of appropriate green supply chain management and logistics principles. More specifically, focus is put on the minimization of the environmental burden and the maximization of supply chain sustainability of the agrifood supply chain. The application of such a framework could result into substantial reduction of CO_2 emissions both by the additional production of other biofuels from waste, as well as the introduction of a novel intelligent logistics network, in order to reduce the harvest and transportation energy input. Moreover, the expansion of the biomass feedstock available for biofuel production can provide adequate support towards avoidance of food/fuel competition for land use.

Life cycle assessment (LCA) studies of carbon footprint (CF) of milk from grass-based farms are usually limited to small numbers of farms (<30) and rarely certified to international standards, e.g. British Standards Institute publicly available specification 2050 (PAS 2050). The goals of O'Brien and colleagues in Chapter 3 were to quantify CF of milk from a large sample of grass-based farms using an accredited PAS 2050 method and to assess the relationships between farm characteristics and CF of milk. Data was collected annually using on-farm surveys, milk processor records and national livestock databases for 171 grass-based Irish dairy farms with information successfully obtained electronically from 124 farms and fed into a cradle to farm-gate LCA model. Greenhouse gas (GHG) emissions were estimated with the LCA model in CO_2 equivalents (CO_2-eq) and allocated economically between dairy farm products, except exported crops. Carbon footprint of milk was estimated by expressing GHG emissions attributed to milk per kilogram of fat and protein-corrected milk (FPCM). The Carbon Trust tested the LCA model for non-conformities with PAS 2050. PAS 2050 certification was achieved when non-conformities were fixed or where the effect of all unresolved non-conformities on CF of milk was $<\pm5$ %. The combined effect of LCA model non-conformities with PAS 2050 on CF of milk was <1 %. Consequently, PAS 2050 accreditation was granted. The mean certified CF of milk from grass-based farms was 1.11 kg of CO_2-eq/kg of FPCM, but varied from 0.87 to 1.72 kg of CO_2-eq/kg of FPCM. Although some farm attributes had stronger relationships with CF of milk than the others, no attribute accounted for the majority of variation between farms. However, CF of milk could be reasonably predicted using N efficiency, the length of the grazing season, milk yield/cow and annual replacement rate ($R^2 = 0.75$). Management changes can be applied simultaneously to improve each of these traits. Thus, grass-based farmers can potentially significantly reduce CF of milk. The certification of an LCA model to PAS 2050 standards for grass-based dairy farms provides a verifiable approach to quantify CF of milk at a farm or national level. The application of the certified model highlighted a wide range between the CF of milk of commercial farms. However, differences between farms' CF of milk were explained by variation in various aspects of farm performance. This implies that improving farm efficiency can mitigate CF of milk.

Readily available inexpensive substrate with high product yield is the key to restoring acetone-butanol-ethanol (ABE) fermentation to economic competitiveness. Lactose-replete cheese whey tends to favor the production of butanol over acetone. In Chapter 4, Ujor and colleagues investigated the fermentability of milk dust powder with high lactose content, for ABE production by *Clostridium acetobutylicum* and *Clostridium beijerinckii*. Both microorganisms produced 7.3 and 5.8 g/L of butanol respectively, with total ABE concentrations of 10.3 and 8.2 g/L, respectively. Compared to fermentation with glucose, fermentation of milk dust powder increased butanol to acetone ratio by 16% and 36% for *C. acetobutylicum* and *C. beijerinckii,* respectively. While these results demonstrate the fermentability of milk dust powder, the physico-chemical properties of milk dust powder appeared to limit sugar utilization, growth and ABE production. Further work aimed at improving the texture of milk dust powder-based medium would likely improve lactose utilization and ABE production.

In Chapter 5, Kim and colleagues conducted a life cycle assessment to determine a baseline for environmental impacts of cheddar and mozzarella cheese consumption. Product loss/waste, as well as consumer transport and storage, is included. The study scope was from cradle-to-grave with particular emphasis on unit operations under the control of typical cheese-processing plants. SimaPro© 7.3 (PRé Consultants, The Netherlands, 2013) was used as the primary modeling software. The ecoinvent life cycle inventory database was used for background unit processes (Frischknecht and Rebitzer, J Cleaner Prod 13(13–14):1337–1343, 2005), modified to incorporate US electricity (EarthShift 2012). Operational data was collected from 17 cheese-manufacturing plants representing 24 % of mozzarella production and 38 % of cheddar production in the USA. Incoming raw milk, cream, or dry milk solids were allocated to coproducts by mass of milk solids. Plant-level engineering assessments of allocation fractions were adopted for major inputs such as electricity, natural gas, and chemicals. Revenue-based allocation was applied for the remaining in-plant processes. Greenhouse gas (GHG) emissions are of significant interest. For cheddar, as sold at retail (63.2 % milk solids), the carbon footprint using the IPCC 2007 factors is 8.60 kg CO_2e/kg cheese consumed with a 95 % confidence interval (CI) of 5.86–12.2 kg CO_2e/kg. For mozzarella, as sold at retail (51.4 % milk solids), the carbon footprint is 7.28 kg CO_2e/kg mozzarella consumed, with a 95 % CI of 5.13–9.89 kg CO_2e/kg. Normalization of the results based on the IMPACT 2002+ life cycle impact assessment (LCIA) framework suggests that nutrient emissions from both the farm and manufacturing facility wastewater treatment represent the most significant relative impacts across multiple environmental midpoint indicators. Raw milk is the major contributor to most impact categories; thus, efforts to reduce milk/cheese loss across the supply chain are important. On-farm mitigation efforts around enteric methane, manure management, phosphorus and nitrogen runoff, and pesticides used on crops and livestock can also significantly reduce impacts. Water-related impacts such as depletion and eutrophication can be considered resource management issues—specifically of water quantity and nutrients. Thus, all opportunities for water conservation should be evaluated, and cheese manufacturers, while not having direct control over crop irrigation, the largest water consumption activity, can investigate the water use efficiency of the milk they procure. The regionalized normalization, based on annual US per capita cheese consumption, showed that eutrophication represents the largest relative impact driven by phosphorus runoff from agricultural fields and emissions associated with whey-processing wastewater. Therefore, incorporating best practices around phosphorous and nitrogen management could yield improvements.

Sheep production in Canada is a small industry in comparison to other livestock systems. Because of the potential for expansion of the sheep industry in Canada, in Chapter 6 Dyer and colleagues assessed the GHG emissions budget of this industry. The GHG emissions from Canadian lamb production were compared with those from the Canadian beef industry using the ULICEES model. The GHG emission intensity of the Canadian lamb industry was 21% higher than lamb production in France and Wales, and 27% higher than northern England. Enteric methane accounts for more than half of the GHG emissions from sheep in Canada. The protein based GHG emission intensity is 60% to 90% higher for sheep than for beef cattle in Canada. The GHG emission intensity for sheep in Eastern Canada is higher than for sheep in Western Canada. Protein based GHG emission intensity is more sensitive to the difference between sheep and beef than LW based emission intensity. This paper demonstrated that protein based GHG emission intensity is a more meaningful indicator for comparing different livestock species than live weight (LW) based GHG emission intensity.

Sustainability marketing trends have typically been led by smaller, more mission-driven firms, but are increasingly attracting larger, more profit-driven firms. Studying the strategies of firms that are moving away from these two poles (i.e., mission-driven but larger firms, and profit-driven firms that are more committed to sustainability) may help us to better understand the potential to resolve tensions between firm size and sustainability goals. In Chapter 7, Howard and Jaffee used this approach to analyze a case study of the U.S. fair trade coffee industry, employing the methods of data visualization and media content analysis. The authors identified three firms that account for the highest proportion of U.S. fair trade coffee purchases (Equal Exchange, Green Mountain Coffee Roasters and Starbucks) and analyzed their strategies, including reactions to recent changes in U.S. fair trade standards. They found an inverse relationship between firm size and demonstrated commitment to sustainability ideals, and the two larger firms were much less likely to acknowledge conflicts between size and sustainability in their public discourse. The paper concludes that similar efforts to increase sustainability marketing for other products and services should be more skeptical of approaches that rely on primarily on the participation of large, profit-driven firms.

Many small-scale farms and several tea estates have started tea plantation in the north-eastern parts of Panchagarh District due to favorable soil and climate. But the conventional approach of tea cultivation using agrochemicals is negatively affecting soil natural fertility; such tea plantation has resulted in significant soil degradation. The productivity has declined and expansion of the industry has threatened by poor conventional management practices. Excessive and unbalanced use of agro-chemicals has led to increase production costs but decline in farm productivity. Thus, there is growing emphasis in the region for ecological and or sustainable approach in tea cultivation to replace the conventional approach. Now soil management in tea gardens by ecological approach is preferred in these areas. Farm scale integrated natural resource management could be a potential solution for tea soil management, while organic sources may reduce the dependency on chemical fertilizers. It is evident from the estimation of fertilizers Chapter 8, by Sultana and colleagues, that chemical fertilizers should be avoided and/or minimized by adoption of integrated natural resource management based on organic sources such as poultry manure, cowdung, farm yard manure and compost. The chemical fertilizer requirements for young tea plants might be avoided by use of poultry manure as they require comparatively low total amount of N and P, while the chemical fertilizer requirements for mature tea can be avoided or minimized by the use of cow dung; almost 50% of N requirement and total amounts of P and K requirements might be possible to meet up. Thus, farm scale resource use efficiency and farm profitability will be sustainable for tea growers in Panchagarh District.

Chapter 9, by Mărculescu and colleagues, presents the results of the experimental analysis oriented to the quantification of some physical and chemical properties of meat processing industry residues and the assessment of their potential as renewable energy source. Experimental laboratory analyses were conducted to establish the elemental composition, volatile fraction, fixed carbon and inert content of the products. The calorimetric experiments conducted on representative waste samples revealed the high and low heating value of the product. The results enabled the identification of possible waste to energy conversion solution. Further investigations are in progress for practical applications.

In Chapter 10, Rada and colleagues describe batch anaerobic digestion tests carried out on stillages, the residue of the distillation process on fruit, in order to contribute to the setting of design parameters for a planned plant. The experimental apparatus was characterized by 3 reactors, each with a useful volume of 5 liters. The different phases of the work carried out were: determining the basic components of the COD of the stillages; determining the specific production of biogas; and estimating the rapidly biodegradable COD contained in the stillages. In particular, the main goal of the anaerobic digestion tests on stillages was to measure the parameters of Specific Gas Production (SGP) and Gas Production Rate (GPR) in reactors in which stillages were being digested using ASBR (Anaerobic Sequencing Batch Reactor) technology. Runs were developed with increasing concentrations of the feed. The optimal loads for obtaining the maximum SGP and GPR values were 8-9 gCOD L^{-1} and 0.9 gCOD $g^{-1}VS$.

Chapter 11, by Jin, is an introduction to a special issue of *AIMS Bioengineering*—"Bioconversion for Renewable Energy and Biomaterials." Food processing industry generates approximately 45% of the total organic pollution as wastewater and solid wastes. These organic pollutants pose increasing disposal and environmental challenges. The treatment and disposal of the organic wastes require many successive and costly treatment processes. These organic pollutants contribute high organic loading in organic carbon and nutrient sources. The processing effluents and solid wastes from food industries mainly contain carbohydrate organics such as sugars, starch and cellulose. They are biodegradable materials and naturally rich in nutrients, making them ideal substrates for microbial production. Most of the existing treatment systems for the food processing wastes worldwide, however, are of old-fashioned processes and cause large losses of valuable nutrient and carbon resources. Considering increasing global concerns due to greenhouse gas emission and resource crisis, there is a general agreement that environmental protection can only be achieved by integrating a general environmental awareness into a company's business functions, making the carbohydrate wastes as renewable resources.

Chapter 12, by Reynolds and colleagues, reviews the current literature around the environmental impacts of dietary recommendations. The focus of the review is on collating evidence relating to environmental impacts of the dietary advice found in the World Health Organisation guidelines, and environmental impact literature: reducing the consumption of fat, reducing the consumption of meat-based protein and animal-based foods, and increasing the consumption of fruit and vegetables. The environmental impact of reducing dietary fat intake is unclear, although reducing consumption of the food category of edible fats and oils appears to have little impact. However most, but not all, studies support environmental benefits of a reduced consumption of animal-based foods and increased consumption of fruit and vegetables. In general, it appears that adhering to dietary guidelines reduces impact on the environment, but further study is required to examine the environmental impacts of animal-based foods, and fruit and vegetable intake in depth.

PART I

OVERVIEW OF FOOD PRODUCTION AND SUPPLY CHAINS

CHAPTER 1

Current Trends in Green Technologies in Food Production and Processing

JOYCE I. BOYE AND YVES ARCAND

1.1 INTRODUCTION

Food is an integral component of life and human existence. Since the beginning of time, humans have had to eat to survive. In earlier times when human population was much smaller, resources were abundant and there was less need for food processing and storage. As populations grew, limitations in food processing and storage techniques forced more individuals to devote considerable amounts of time daily to feeding themselves and their families (i.e., harvesting and hunting). Industrialization shifted a large percentage of the population toward a myriad of activities creating the need for an industrialized food sector to feed an increasing number of urbanized humans.

Burgeoning population growth in the last several centuries with the attendant demands resulting from industrialization has made the need for sustainable food production and processing technologies even more im-

Current Trends in Green Technologies in Food Production and Processing. © *Boye JI and Arcand Y. Food Engineering Reviews 5,1 (2013). doi: 10.1007/s12393-012-9062-z. Licensed under Creative Commons Attribution License, http://creativecommons.org/licenses/by/3.0/.*

portant. At the same time, changes to climate and population health have made evident the precarious balance between sustainable food production practices, a healthy environment, and a healthy population. In 2050, the population of the world is expected to reach 9 billion FAOSTAT [23]. Adequate supplies of healthy, nutritious food will be needed to maintain global socioeconomic viability. To do this successfully will require that we produce more food with much less impact on our environment. Our ability to meet growing demands for food supplies will, thus, hinge on the sustainability of the practices used in food production and the fervor with which novel processes and technologies are developed to address ever-changing and conflicting pressures.

Since the industrial revolution, worldwide food production has increased significantly but at a slower pace than global population and with much more waste and less efficient resource distribution. Food supply shortages have left 3 billion people malnourished globally with iron deficiency affecting 2 billion people and protein/calorie deficiencies affecting nearly 800 million people [25]. At the same time, most land and aquatic resources are overused. And even more startling is the estimate that currently 30–50 % of food produced is wasted [4].

Among some of the more serious environmental concerns we face are land use change and extensive reduction in biodiversity, aquatic eutrophication by nitrogenous substances caused by over-fertilization, global warming caused by enteric fermentation and use of fossil fuels, aquatic eutrophication by phosphorous substances caused by fertilizers overuse, water shortages due to irrigation, ecotoxicity, and human effects of pesticides [7].

Growing awareness of these challenges is causing social shifts with some stakeholders including farmers, food manufacturers, consumers, and policy makers, desiring more efficient approaches in agricultural and food production practices. Increasing use of organic inputs in processing, use of recyclable and good-for-the-environment packaging, establishing of just employer–employee relationships, "buy local," "whole foods," "free-from," and "fair-trade" are examples of some of these trends.

Green technology is defined by the global collaborative encyclopedia, Wikipedia, as "the application of one or more of environmental science, green chemistry, environmental monitoring and electronic devices to monitor, model and conserve the natural environment and resources, and to curb the negative impacts of human involvement" (http://en.wikipedia. org). In the field of agriculture and agri-food, the term "green growth" is sometimes used and has been defined by the Organization for Economic Co-operation and Development (OECD) as "the pursuit of economic growth and development, while preventing environmental degradation, biodiversity loss and unsustainable natural resource use" [41]. This review summarizes key highlights from the recently published book entitled *Green Technologies in Food Production and Processing* [7]. Issues addressed include key drivers of the evolution in the food supply chain, in-depth description of food production, and processing using the life cycle assessment (LCA) tool; approaches to improve food production practices; sustainable food processing approaches; emerging analytical techniques for sustainable research and development; challenges associated with the use of agricultural resources to grow biofuels and bio-based products; technologies to reduce the generation of process-induced toxins; social factors that influence consumer perceptions about some of the current and emerging agri-food technologies; and the need and importance of biodiversity in maintaining sustainable diets of human populations.

1.2 THE FOOD CHAIN

Approximately 10,000 years ago, the beginning of agriculture contributed to an intense transformation in the organization of social life, which evolved from the hunter-gatherer and nomadic societal structure to sedentariness. The food chain was very simple—as conservation techniques were limited, foods were grown locally and eaten only when in season. A subsistence agricultural system prevailed, and because of its importance for survival, food was not considered as simply another commodity but an important requirement for existence [28].

The subsistence model was well suited to rural areas or small towns. With the rapid growth of industrialization came larger cities which distanced farmers from urban workers. Retailers appeared as intermediaries between producers and consumers. Urban workers labored long hours to increase their standard of living and consequently demanded more sophisticated foods while having less time for food preparation at home. Food processors were grafted into the food chain to address the gap and provide consumers with partially prepared as well as ready-to-eat (RTE) meals.

During the second half of the twentieth century, key stakeholders in the food chain (producers, processors, distributors, and retailers) increased in size, becoming large industries with a desire to sell large offerings of very specialized products to a global market. This culminated in the signing of the important World Trade Organization multilateral trade agreements in the 1990s. Since then, trade in agricultural products has increased tremendously, from 280 billion US$ in 1999 to 920 billion US$ in 2009 [26]. Today, it is not uncommon to have similar products from different countries available at a given grocery store or to have what used to be seasonal produce available all year round because suppliers come from both hemispheres.

A supply chain can be defined as "a system whose constituent parts include material suppliers, production facilities, distribution services and customers linked together by a feed-forward flow of material and feedback of information" [51]. The agri-food supply chain in particular refers to a system of actors linked from "farm to fork" to produce consumer-oriented products in a more effective manner and with an optimized flow of agricultural products through the different steps of the chain [28]. The food chain has some specific issues such as (1) seasonality of supply and demand, (2) customer issues of traceability and risk management related to health, nutrition, and safety, and (3) the environmental impact of food production through extensive resource use, including water and land use and from the significant greenhouse gas emissions and waste resulting from agricultural production [36].

It is not surprising that the food chain has become more complex as the income of the average population has risen. Low-income populations eat what they can afford but, as a population gets wealthier, the criteria for food selection increases and food choice is increasingly affected by factors such as food safety, good taste, long shelf life, non-GMO, and additional health benefits. Table 1 summarizes the main differences in the supply chain of low-, medium-, and high-income countries.

TABLE 1: Main characteristics in the supply chain for low-, medium-, and high-income countries

Constituents of the food supply chain	Traditional agriculture	Modernizing agriculture	Industrialized agriculture
Inputs intensity	Low input use	High level of use	Enhanced input use efficiency
Primary agriculture	Diversified	Specialization of cropping systems	Specialization and focus on conservation
Processing sector	Very limited	Processed products are seen as value-added and provide employment	Large processing sectors for domestic and export markets
Wholesalers	Traditional wholesalers with retailers bypassing for exports	Traditional and specialized wholesalers	Specialized wholesalers and distribution centers
Retailers	Small market	Spread of supermarkets, less penetration for fruits and vegetables	Widespread supermarkets.
Consumers	Rising caloric intake	Diet diversification, switch to processed foods	High value, processed foods.
Traceability	No traceability	In some chains with private standards	HACCP programs

Source: [26]

A significant trend in the structure of agri-food supply chains is the growing concentration and consolidation that has taken place at all levels of the chain. This concentration is a concern if it results in collusion, price fixing, and unfair market practices. Competition policy exists in most developed countries but there is no doubt that the activities controlled by large firms may have significant implications for the structure and management of the supply chain [36].

At the farm level, the race for higher productivities has resulted in the use of bioengineered varieties and more sophisticated inputs (e.g., machinery, fertilizers, and pesticides). In some instances, the monopoly of the seed supply has generated monocultures of only one variety which can have a large impact on biodiversity.

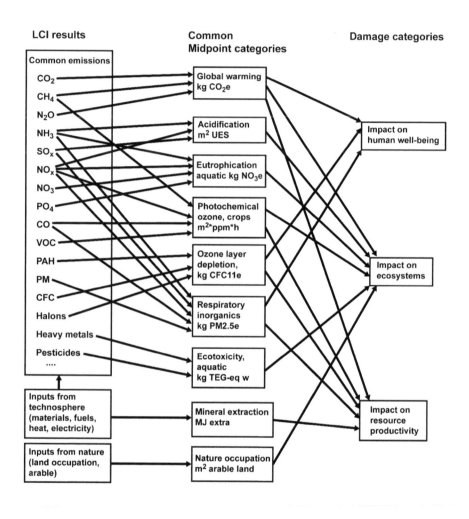

FIGURE 1: General structure of an LCA model linking inputs and outputs of inventory to midpoint and final impacts. Source: [32]

Primary producers, food processors, and retailers are under constant pressure from domestic and international markets to improve efficiency of their operations and reduce costs while satisfying customers. Most large companies have introduced into their quality control systems some form of traceability not only internally but also with suppliers. This includes not only product quality and safety but also some environmental considerations such as energy reduction, water issues, waste reduction, packaging reduction, greenhouse gas (GHG) emission, transportation, sustainable agriculture, renewable energy, and carbon footprint. Nestlé and Wal-Mart are two examples of such companies with a large enough market share to have substantial impacts on the sector.

1.3 LIFE CYCLE/ENVIRONMENTAL IMPACT ASSESSMENT

Since the Industrial Revolution, humanity has been living under increasing environmental credit: natural resources are shrinking exponentially while wastes are accumulating globally at near parallel rates. Not surprisingly, since the 1970s, there has been a growing outcry about the danger of human activities to its own survival. And yet, today, questions about how to do things differently remain. The apparent immediate solution is the establishment of mitigation measures but, more often than not, these measures simply move the pollution from one place to another (e.g., gas scrubbing).

At the Rio Summit in 1992, sustainable development was defined as: "…development that meets the needs of the present without compromising the ability of future generations to meet their own needs" [11]. For this to be of practical use, tools are needed to measure and compare the environmental impacts of various human activities relative to "sustainable standards."

1.4 LCA OF THE FOOD SECTOR

Life Cycle Assessment (LCA) may be defined as a product-oriented environmental tool which provides a systematic way to quantify the environmental effects of individual products or services from cradle to grave [32]. LCA starts by defining an activity (product or service) and decomposing it into a sequence of sub-processes that use resources to produce (a) desired products, (b) by-products, and (c) unwanted outputs (wastes). By subdividing sub-processes into their smallest elements, it becomes possible to track and tally (embed) the resources used and the unwanted outputs of each elemental sub-process into the complete product (or service) under examination. This is called the Life Cycle Inventory (LCI).

The next step, which is the Life Cycle Impact Assessment (LCIA), estimates the impacts of resource depletion and waste generation using a series of models. The main strength of LCA is that it tries to have a holistic view of the activity under study by taking all impacts into account and fusing all models into one coherent model. Figure 1 shows one such supra-model. As shown, production of one portion of meat from beef generates a variety of emissions. The model shows numerous midpoint impacts and thus illustrates how this activity can alter resource productivity, ecosystems, and human well-being.

In LCA, final impact results are easier to interpret than midpoint impacts. But, conversely, the accuracy of the final results is weaker because the uncertainty associated with the inventory data is increased by the uncertainty of the different impact models. LCA is, nevertheless, now well recognized internationally and has its own ISO standards which define the general methodology, variants, and accuracy of the different results (ISO 14 040–14 049, European Commission—Joint Research Centre—Institute for Environment and Sustainability [17]. As indicated in the ISO standards, conclusions depend strongly on the chosen assumptions. First and foremost is the choice of the type of LCA (attributional vs consequential). The attributional approach describes the resource use and emissions that have occurred to produce the product in question, whereas the consequential approach accounts for the resource use and associated emissions that arise to replenish stocks of the product that have been used [32].

Another set of assumptions comes from the definition of the functional unit and the boundaries of the system under study. An example of an ill-defined functional unit is "1 kg of beef meat packaged at the retail store in Toronto." A well-defined functional unit will, at least, say which race of beef, where it was raised, what its diet was, age of animal when killed, how and where it was slaughtered, how it was transported, what kind of packaging was used, and so on. Moreover, allocations are often necessary to adequately associate specific emissions to different products and by-products. For example, cows provide milk (during their life) and meat (upon slaughter). This raises the question of which proportion of the methane produced every day should be attributed to each liter of milk versus each kg of meat produced. General rules in setting these assumptions are constantly revised to ease the comparison between different LCA results.

LCA is a generic tool that is adaptable to any situation and which is evolving quickly. The impact of land availability, as an example, has been added as a resource to get a fair comparison between intensive and extensive agriculture. Intensive agriculture uses more fertilizers, whereas extensive agriculture has larger impacts on deforestation and loss of biodiversity. Modeling of these resources and impacts is in its infancy and is changing rapidly.

1.5 LCA OF FOOD FROM PLANT ORIGIN

The compatibility of agricultural productivity and sustainability is an important socioeconomic question [10]. Foods of plant origin (e.g., fruits, vegetables, and cereals) are produced through gathering either from natural habitats or from man-made plantations. Man-made cultures started over 10,000 years ago when populations increased to the point where gathering within walking distance became impossible. Today, human populations have reached a point where large proportions of the world's natural habitats have been transformed for agriculture. Between the years 2000 and 2030, the world population is projected to increase by 40 %, cereal production by 50 % (due to improved standard of living in certain countries), whereas arable land is expected to increase by only 7 % FAO [20, 21]. This will require agriculture to be more efficient (more intensive) which will put more pressure on rural areas. Plants need minerals, which come from the soil, to grow. To ensure productivity and profitability, minerals have to be added in slight excess without affecting the surroundings. LCA provides a suitable tool to quantify the resources needed, the wastes generated, and the impacts of agricultural activities. Numerous studies have shown the significant effect of nitrogen and phosphorus fertilization on eutrophication, acidification, global warming, and photochemical ozone. Additionally, particulate formation (dusts), loss of biodiversity caused by land use change, toxicity, and depletion of abiotic resource (it takes crude oil to produce the fertilizers) have raised concerns [10].

For crops, a functional unit is generally considered as a ton of grain ready to leave the farm. Contrary to other sectors, agricultural production often takes place within an open space (i.e., without distinct borderlines for soil, water, and air) and with uncontrolled conditions (temperature, rainfall, extreme weather). This adds a considerable amount of uncertainty to nutrient runoff or volatilization measurements or estimations for modeling which can change the conclusion of any LCA tremendously. This explains why recent studies tend to use average multiyear values under standardized conditions [5, 8, 53].

Another specific element of many agricultural LCIs is the need for land use data. Former quantification data sets were limited to the area used per functional unit but more recent studies tend to add the quality of the original area into account as transforming a desert into agricultural land does not have the same impact as using a rainforest for the same purpose Brentrup et al. [9]. Besides these direct effects, indirect land use impact may also be of relevance (e.g., if the cultivation of bioenergy crops displaces food crops to other areas).

1.6 LCA OF FOOD OF ANIMAL ORIGIN

The world population increased exponentially to reach 7 billion in 2011 and the majority of the projected increase to 9 billion by 2045 is expected to come from Asia and Africa FAOSTAT [23]. At the same time, rural populations are decreasing in every region of the world to the point where some rural populations may disappear in the next 30 years. Global growth rates of animal product consumption (meat, milk and eggs) per capita is higher in wealthier countries and increases more rapidly in countries with higher economic growth rates FAO [22]. Although there are differences between regions, meat consumption per capita will globally increase by 7 % in developed countries, 33 % in Latin America, 65 % in the Middle East and North Africa, 82 % in Eastern, Asia and 100 % in sub-Saharan Africa. To meet this growing demand, animal herd has increased by a factor of 2, 2.5, and 5 for beef, swine, and poultry, respectively FAOSTAT [23]. For proteins derived from sea products, the catch has increased 6.5-fold in the last 50 years WHO/FAO [59].

Livestock production systems have evolved in past decades due to increasing demand for animal protein, on one hand, and economic growth rates and technical innovations, on the other. Over time, local family farms were sold or transformed into sophisticated high production factories. This led to the adoption of high-energy diets, genetic improvements, high-density animal concentrations, and the use of growth hormones and antibiotics. Globalization also concentrated these animal factories in areas where the economics are favorable without due consideration of sustainability.

LCA of animal production involves mainly feed, water, and land use. Water and land use for the sole production of the animal is not very important but the water and land use embodied into the production of feeds are significantly more important, leading to two different production systems: extensive and intensive [56]. In extensive production systems, farmers produce the feed needed locally. The number of animals in such systems is restricted in order not to exceed the grass or crop feed capacity of the land and in so doing minimize the risk of soil degradation [14, 16, 60] or deforestation [18, 19]. In intensive production systems, animals are more densely packed into barns and the feed is purchased from specialized companies preparing high-energy rations from grains that may be located anywhere in the world. This "distanciation" between crop and manure production forces the injection of fertilizers to the crop land and the disposal of excess manure to nearby fields which could lead to water and air pollution [37].

Moreover, the use of cereals for animal feeding represents about half of global cereal supply [39]. As the land required for one portion of animal protein can give close to 10 portions of plant proteins, it is not surprising that the land used for animal feeding has become an important issue.

Animal production affects biodiversity in a variety of ways including through (a) soil degradation in overused extensive systems, (b) land use change to crop monocultures, and (c) animal selection [2]. Furthermore, the effect of animal production on global warming has become a matter of increasing concern. Carbon dioxide (CO_2), methane (CH_4), and nitrous oxide (N_2O) are the three main contributors to GHG emissions. They come mainly from enteric fermentation, from manure and fertilizers degradation as well as in association with biological nitrogen fixation in pulse and legume crops [56]. The global warming potential of methane and nitrous oxide is about 20 and 300 times the potential of carbon dioxide. In 2005, global agriculture was estimated to account for about 11 % of all anthropogenic GHG emissions. The agricultural share of total anthropogenic CH_4 and N_2O emissions was about 47 and 58 %, respectively [49]. Water required for animal production is generally higher than that required for crops because the direct water consumed by the animal and the water used to produce its feed have to be taken into account [50]. The world average water footprint per kg of meat was estimated at 15,500 L for beef, 6,100 L for sheep, 4,800 L for pork, and 3,900 L for poultry [33]. Water runoff from animal production may also be of concern because of its high capacity to pollute receiving water surface or groundwater [35].

Vergé et al. [56] in reviewing the production of milk, beef, pork, poultry, and aquaculture and their impacts on global warming concluded that comparing results of LCAs is very difficult because of the different assumptions in the type of LCA, the functional unit, the system boundaries, and method of allocation chosen. Notwithstanding these differences, studies indicate a net positive impact of mitigation practices and beneficial management practices.

1.7 LCA OF FOOD PROCESSING

It is generally accepted that the most important "contributors" to the non-sustainability of food production (i.e., along the whole food chain) are pesticide overuse, food processing, packaging, and extensive transportation of imported goods. Extending the LCA to include food processors therefore seems judicious. In fact, the first "LCA inspired" studies date back to the 1960s and come from food processors [34] analyzing the effect of different packaging (glass vs. plastic soda bottles and polystyrene foam vs. paper pulp meat trays).

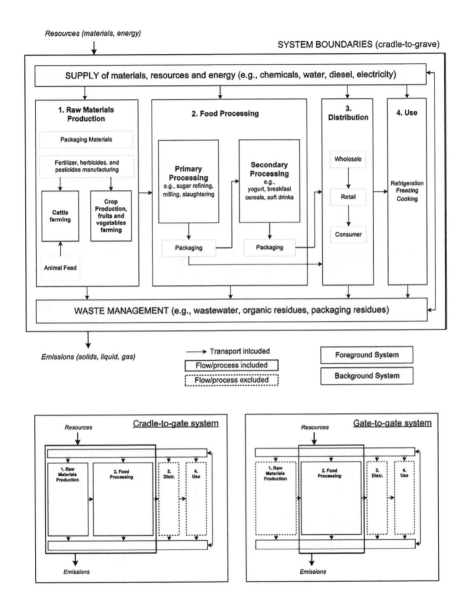

FIGURE 2: System boundaries for food LCA study. Cradle-to-grave (top), cradle-to-gate (bottom left), and gate-to-gate (bottom right) system boundaries. Source: [3]

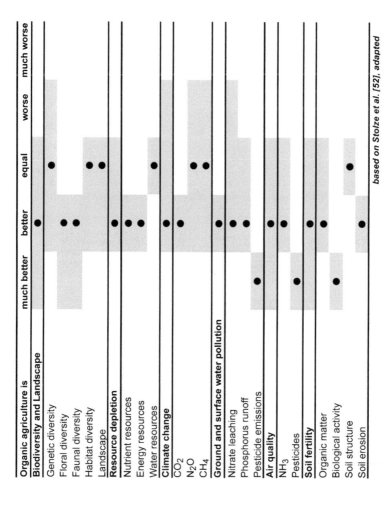

FIGURE 3: Classification of environmental impacts and relative performance of organic farming compared to conventional farming Source: Schader et al. [44]

Food processing is a diversified sector encompassing the use of various raw materials, processes, and end products. It ranges from minimal (e.g., selling fresh carrots) to very complex (e.g., carrots that are washed, cut, and packaged to be sold to food processors making sauces and stews who sell to caterers preparing individual sized meals to be sold to supermarkets or for airline catering).

LCA of processors are "extensions" of cradle-to-farm-gate LCAs (Fig. 2). Boundary expansion is generally easy to do because the boundary for food processing extends only as far as the physical limit of the manufacturing facility. However, the processor's functional unit is generally very different from the farmer's which leads to difficulty in allocations. A simple example is the proportion of the GHG produced by a cow (farmer's functional unit) that should go to each hectoliter of milk sold (dairy processor unit) versus to the carcass sold (meat processor unit)? A more complex example would be to assess how much GHG can be tagged to a meal of beef stew eaten with a sauce containing milk solids and a cream based dessert?

Raw materials used in the food and drink industry are of agricultural origin and are made from limited resources of land, water, and energy and produced using manufactured inputs such as fertilizers, pesticides, and cleaning agents. Environmental issues related to the food processing chain mainly include reduction in biodiversity (i.e., through land use change and monocultures); wastewater generation (including biological oxygen demand) and nitrogen, phosphorus, and sulfur availability; particles and toxicity; and generation of packaging wastes and organic residues (including by-products production) as well as air pollutants emissions (i.e., volatile organic compounds, particulate matter, greenhouses gases, odors, and refrigerant leaks). Non-environmental issues include diet simplification, animal welfare, workers' conditions, and the generation of acceptable social and economic conditions. The relative importance of these factors depends strongly on the food sector and the stakeholder within the food chain but, in general, food processing has lower environmental impacts than farming [3]. Either because they are forced to by legislations or as a direct result of their own initiatives, an increasing number of food processors are evaluating their business activities in order to report, improve, and/or market their environmental efforts including their supply chain. Proactive initiatives are driven, on one hand, by the positive image processors and retailers can convey to purchasers and consumers. On the other hand, performing an LCA reveals potential areas to reduce inputs, energy, and water which has cost savings advantages and consequent environmental benefits.

1.8 LCA OF TRANSPORTATION ALONG THE FOOD CHAIN

As food products often travel long distances to reach consumers, one might expect transportation to be the major food-related contributor to greenhouse gas emissions in many developed countries. In fact, almost half of the fruits consumed in North America is imported and domestic grown produce travel an average of 2,000 km from source to point of sale, not infrequently via air [42]. Movement of food is done through a highly complex, relatively integrated industry sector including ships, trains, trucks, planes, and warehouses and often require special packaging [57]. Food supply chains are especially challenging because of seasonality, freshness, spoilage, and sanitary considerations.

Weber and Matthews [58] showed that international water shipments have the lowest GHG impact closely followed by inland water and rail; trucking produces 10 times more GHG per ton-km travelled, whereas air transportation is 40 times higher. Unfortunately, faster modes of transport produce more GHG per ton-km. Consequently, the impact of transportation depends on the type of transportation mode, the delay from source to consumer and the transportation conditions (i.e., temperature, special atmosphere, and frailness). Most LCA studies present results of transportation as a percentage of the impact of the total food chain. Depending on products and processes, the impact of transportation on GHG ranges from less than 5 % to more than 50 % (lower percentage values are generally found with products already embedding large GHG from farming and processing (e.g., milk powder)).

1.9 GREEN TECHNOLOGIES IN FOOD PRODUCTION

Green food production often evokes organic farming practices typical of a few centuries ago. This type of farming uses a small area of land for crops and another dedicated area for grazing beef, sheep, and goat. Farm entities were almost always self-sufficient with no use of pesticides or herbicides and the only fertilizer used was manure. As a typical cradle-to-cradle approach, organic farming suits the notion of a green technology. As bucolic as the image is, these farming practices have to be evaluated for sustainability. A generally valid quantification of the environmental performance of organic agriculture is, however, difficult as there is high variability between countries, regions, farm types, and products. Furthermore, different assessment methods lead to partly contradicting conclusions on the environmental impacts of organic farming. Organic farming performs better in terms of biodiversity, soil fertility, air quality, mitigating resource depletion and climate change, and groundwater pollution as compared to conventional agriculture [44]. However, there are specific environmental impacts against which organic agriculture performs no better than conventional farming (Fig. 3).

A more specific aspect of green food production is the management of nitrogen and phosphorus. These nutrients are taken up from the soil by crops and returned partially through manure management. In intensive production, these nutrients rapidly become the growth limiting factor. Fertilizers must be applied in slight excess to be certain crops will not be growth limited. In highly specialized production food systems, crop growers do not own animals and thus manure is not readily available. Crop producers must rely on external inputs of fast release mineral nutrients. These products undergo numerous chemical transformations depending on the type of soil, type of crop, period of the year, and the immediate weather. Unfortunately, this often leads to nitrogen volatilization and nitrogen and phosphorus runoff during rainy weather. On the other hand, concentration of animal production produces a surplus of manure requiring disposal. This often results in manure overspreading which leads to odors and nutrients percolating into the aquifer or runoff into surface water. Correctly managing the nitrogen and phosphorus agronomic cycles can help regain soil fertility and minimize pollution. Vayssières and Rufino [55] developed a detailed model of the nitrogen cycle on farm of different farming management practices and showed that the sources of nitrogen losses within the nutrient cycle in agro-ecosystems are numerous and the losses substantial. As an example, estimates suggest that about 15 % of N used in mineral fertilizers and 30 % of N excreted by domestic animals worldwide are globally lost in the form of harmful gases (NH_3 and N_2O). A holistic approach along the production value chain is needed to manage these losses and minimize negative impacts.

As most of the demand for food products in the near future will come from Africa and southeast Asia, it is important to identify affordable management practices for these countries. With a well-chosen combination of technical options, N-use efficiency can be substantially improved in farming systems of both developing and industrialized countries.

TABLE 2: Emerging technologies for microbial control in food processing

Technologies	Examples
A) Biopreservation	Bacteriocins
	Organic acids
	Probiotics
B) Electromagnetic wave heating	Microwave technology
	Radiofrequency technology
C) Electric and magnetic fields	Ohmic heating
	Moderate electric field heating
	Inductive heating
D) Nonthermal technologies	Pulsed electric field
	High pressure processing
	Ionizing radiation
	Ultraviolet radiation
	High-intensity pulsed light
	Ultrasound
	Ozonization
	Cold plasma processing

Source: [40]

1.10 GREEN TECHNOLOGIES IN FOOD PROCESSING

Foods must frequently be processed to ensure safety and increase shelf life, quality, and nutritional properties while making them more convenient. Primary, secondary, and tertiary processing techniques are explored to transform raw produce into value-added foods and ingredients. Primary processing techniques such as cleaning, sorting, grading, dehulling, and milling are used as a first step in the processing of most grains. In the animal sector, primary processing includes but is not limited to, candling and harvesting of eggs; deheading, gutting, filleting, scaling, washing, chilling, freezing, or packing of fish; storage, separation, homogenization, and pasteurization of milk; slaughter, dressing, boning acidification, salting, brining, smoking, thermal processing, refrigeration, and storage of meat. Secondary and tertiary processing techniques are further applied to transform these foods into other value-added food products.

One of the most promising technological approaches to reduce environmental footprint in food processing is the use of enzymes. As biological catalysts, enzymes speed up reaction rates and in so doing offer savings in terms of time, energy, and cost. Food enzymes provide advantages in terms of specificity, sensitivity, their relative non-toxicity, high activity at low concentrations, and ease of inactivation. Enzymatic approaches entail milder treatments and/or mild reaction conditions, thus are more environmental friendly and would protect the environment better compared to traditional methods [48]. Furthermore, the discriminatory nature of enzymes in terms of the specificity of molecules they act on as substrates results in more uniform and more consistent products. In addition to environmental benefits, the action of enzymes in foods may result in products with extended shelf life, improved textures, appearance, flavors, functionality, and yield, enabling a variety of food products to be fabricated from harvested produce [48]. Examples of enzymes that can be used in food processing include carbohydrases (e.g., amylases, pectinases, cellulases, galactosidases, and chitinases); lipases (e.g., pancreatic lipase and phospholipases); proteases (e.g., pepsins, trypsins, bromelain, papain, amylases, and cellulases); isomerases (e.g., glucose isomerase); transferases (e.g., transglutaminases); and oxidoreductases (e.g., glucose oxidase and polyphenol oxidase).

Trends in enzyme-assisted food production include enzyme engineering aimed at developing enzymes with superior activities which can be used under mild processing conditions (e.g., nonthermal food processing operations) or which can resist extreme conditions of pH, temperature, and pressure encountered during food processing. Technologies for increasing the yield and storage stability of these enzymes are also of interest. Furthermore, opportunities exist in engineering design to develop application formats that enhance ease of use, recovery, and reuse.

Another priority concern in food processing is food safety. Thermal treatments such as pasteurization, sterilization, aseptic processing, refrigeration, and chemical preservatives have been traditionally used to decrease microbial loads in foods and to enhance safety and shelf life. As many of these operations are energy intensive, alternative techniques that require less energy and that have less environmental impacts should be considered. Novel and innovative methods of microbial control in food processing include microwave and radio-frequency heating (MW/RF), pulsed electric fields (PEF), high pressure processing (HPP), ionizing radiation, ohmic heating (OH), treatment with ultraviolet light, and ozonisation [40]. Table 2 provides a list of some of the alternative techniques that could be considered for microbial control.

Drying, which represents a significant cost investment and a major source of energy consumption for most companies, is another important unit operation that must be considered when a greener process is targeted. Drying is a critical unit operation in the processing of many bulk and packaged food products and ingredients. It is used to provide texture, enhance shelf life, and decrease transportation costs. The typical elements of a drying operation include wet feed pre-treatment, drying, retrieval of dried product, and heat recovery from exhaust gases. To reduce energy consumption, elimination of the drying operation from the production process altogether, or its replacement with lower-energy consuming operations, should be envisaged [29]. Furthermore, whenever possible, initial moisture content of the wet feed to be processed should be reduced using less energy-intensive techniques such as pressing, membrane separation, filtration, centrifugation, coagulation, and sedimentation, prior to the drying process. As an example, osmotic dehydration can be considered as a preheat treatment unit operation or a final dehydration step [30].

Preheating of wet feed to as high a temperature as possible using energy-efficient means can also help to reduce overall energy use. Additionally, utilization of environmentally friendly energy sources and energy-efficient drying installations and maximal use or recycling of different waste streams and by-products are useful considerations [29]. Furthermore, to reduce or eliminate environmental pollution, efficient installations which can completely recover energy, particulates, and greenhouse gases from exhaust gases must be considered.

TABLE 3: Energy and thermal efficiency of selected industrial dryers

Method or dryer type	Energy or thermal efficiency (%)
Tray, batch	85
Tunnel	35–40
Spray	50
Tower	20–40
Flash	50–75
Conveyor	40–60
Fluidized bed, standard	40–80
Vibrated fluidized bed	56–80
Pulsed fluidized bed	65–80
Sheeting	50–90
Drum, indirect heating	85
Rotary, indirect heating	75–90
Rotary, direct heated	40–70
Cylinder dryer	90–92
Vacuum rotary	Up to 70
Infrared	30–60
Dielectric	60
Freeze	10 or lower

Source: [29]

The typical elements of a green drying installation scheme are presented in Fig. 4. To reduce energy use and environmental impacts, heat is recovered from exhaust gases and is recirculated in the drying operation. Exhaust gases after heat recovery can be further scrubbed to remove greenhouse gases. Whether for drying or for other unit operations, the type of energy source used in food processing can have an impact on the total amount of energy consumed and the environmental footprint.

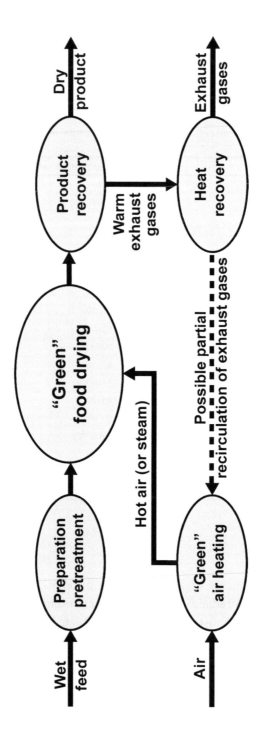

FIGURE 4: Typical elements of a "green" drying installation scheme. Source: [29]

There is general understanding of the need to move from carbon-based energy sources to solar, hydroelectric, and wind. Generally, solar and wind energy sources are natural, abundant, and considered clean and environment-friendly as they create no, or minimal, greenhouse gases. Hydroelectricity is considered by some to be not as environmentally friendly as solar or wind because it involves flooding large area of land which produces methane through anaerobic degradation of the organic material flooded for decades. Debate continues over the safety of nuclear energy and further considerations are needed on the risks versus benefits in terms of health and environmental impacts. A comparison of the energy and thermal efficiency of selected industrial dryers is provided in Table 3.

A final unit operation in many food processing plants prior to storage and distribution is packaging. This is an important activity that has impacts on product acceptability, choice, safety, and nutritional quality. Packaging also provides a means of communication with consumers and allows foods to be portioned in convenient formats. Under-packaging puts foods at risk, whereas over-packaging has high environmental footprint. Greener packaging design considerations should include the maintenance of required functionality, material use minimization, increasing recycled content and use of recyclable materials, and avoidance of potentially toxic constituents [46]. As for other operations, LCA could be a useful tool to evaluate the benefits and disadvantages of alternative packaging systems.

1.11 REDUCING PROCESS-INDUCED TOXINS IN FOODS

Many food processing treatments use high temperatures and sometimes pressures (e.g., extrusion cooking) which can generate process-induced toxins. Examples of toxic compounds of concern in foods include nitrosamines, heterocyclic aromatic amines, acrylamide, furans, polyaromatic hydrocarbons, and bisphenol [1]. In addition to reducing the environmental footprint and the use of chemical additives, green technologies in food processing should aim to limit the levels of toxins in foods.

Formation of process-induced toxins and their levels in foods is frequently dependent on the levels of their precursors in foods as well as processing conditions (e.g., method of cooking, duration, and temperature).

Formulation (i.e., ingredient selection), breeding (e.g., potatoes with low levels of asparagine and sugars to decrease acrylamide content in foods), and use of modified cooking conditions and/or alternative cooking methods (e.g., microwave) are examples of approaches that could be considered to minimize the generation of process-induced toxins. Transfer of toxic chemicals from food processing utensils and containers (e.g., BPA and phthalates used in modern plastics) can also occur; thus, appropriate equipment and material selection must be done to avoid the inadvertent presence of these chemicals in foods.

As consumers become increasingly aware of the nutritional quality of the foods they consume, national and international regulatory agencies will need to keep pace through the development of rapid and sensitive detection methods and generation of evidence-based data to assist in the establishment of tolerance and safe levels of chemical toxins in foods.

1.12 WASTE REDUCTION ALONG THE FOOD SUPPLY CHAIN

Paradoxically, superior plants and animals have evolved to have the least efficient nutritious cycle possible. Strangely, this is one of nature's successes as it promotes biodiversity. As an example, when a bird eats a cherry, it will most probably take just a few bites and leave the rest including the stone. When the bird is killed by another animal, the predator feeds a lot less than its subsequent predator and when humans kill the predator, they also feed on a fraction and excrete a non-digestible "waste." All the remains serve as food to other species such as animals, insects, microorganisms, and finally other plants. Thus, nature recycles its main building blocks and permits evolution.

As the human race has mastered agriculture, a large portion of food is thrown away through farming, processing, distribution, retail, consumption, and excretion. Feeding the world, therefore, generates a tremendous amount of waste. Fortunately, most of the waste is biodegradable and is easily incorporated into other nutrient cycles.

Humans discovered quite early that wastes from plants and animals can cause serious diseases and as a result disposed of them in recessed areas or buried them in pits. This was the first efficient waste disposal approach. With the explosion of the human population, food production and processing has become intensive and the amount of waste generated is less easy to dispose of, especially in urbanized communities. Wastes therefore have to be minimized and processed properly throughout the entire food chain in order to ensure the sustainability of the sector.

TABLE 4: Weight proportion of waste, by type, arising from UK food and beverage supply chain from processor to the consumer

Supply chain stage	Food (%)	Packaging (%)	Other (%)	Total (%)
Manufacturing	6.6	1.0	5.2	12.8
Distribution	0.01	0.22	0.02	0.3
Retail	0.9	2.7	0.1	3.7
Household	21.3	9.2	52.7	83.2
Total	29	13	58	100

Source: www.wrap.org.uk, 2012

Wastes occur across the food chain, and while some waste generation may be avoidable, others are not. For example, of the entire corn plant, only the grains are eaten as food; unfortunately, over 95 % waste is unavoidably generated in the process. Although plant breeding techniques have succeeded in creating plants with larger quantities of edible parts and genetic engineering promises to push this further, mankind will never be able to eat 100 % of all the plants and animals we feed on. Other wastes, on the other hand, are avoidable. Examples include, the choice to eat parts of plants and animals we are culturally not used to consuming; accepting to buy "less than perfect" fruits and vegetables; managing food reserves in such a way that we buy and prepare just what we need on a per meal basis.

Food is an important part of human wastes. Data from the UK food and beverage supply chain from the processor to the consumer shows that approximately 40 % (by weight) of all solid wastes involve food and its packaging (Table 4). The table also shows that over 80 % of all wastes come from the consumer. Food waste prevention is a multilevel hierarchy process inspired from the 3RVE (i.e., reduce, reuse, recycle, valorize, and eliminate) waste reduction strategy. One should first try to REDUCE as much as possible and then try to REUSE, and after that try to RECYCLE, then VALORIZE, and if nothing else works, ELIMINATE. The United States Environmental Protection Agency (USEPA) has proposed the hierarchical solution presented in Table 5.

Source reduction starts at the consumer level focusing on behaviors at home and in the hospitality industry (i.e., restaurants, hotels, and catering services). As change in the way customers behave at home may be very difficult, one of the few potential solutions is through education. This may involve learning how to plan meals, cooking just enough, not overfilling our plates, and better recipes to cook leftovers and not so fresh food material. Whereas numerous small initiatives exist for the reduction and reuse of food in the hospitality industry, these are often limited due to fears of lawsuits for poisoning attempts and food safety risks.

Reduction in wastes at the retail level should include initiatives to market different grades of fresh produce or outdated material at different price levels. Initiatives at the wholesale and distribution level should involve just in time victualing and appropriate packaging and temperature control to minimize losses, although these latter two are not important contributors (see Table 4). The Table also shows that food processors are the second largest contributor to food waste. As much of the wastes are neither food nor packaging, the potential for food reduction is not high. Nevertheless, process audits can often lead to optimization which could help in identifying plant specific solutions.

Waste treatment is used when food waste disposal is unavoidable. As most wastes are solid in nature, there are two principal methods available: landfilling and incineration. Landfilling has been used for centuries and is still the cheapest method (although prices are rising rapidly). This solution, however, requires large land areas and may contribute to the production of greenhouse gases (mainly from methane generation) as the food decomposes anaerobically. This method is suitable for both rich and poor countries. Burning of food waste requires a much smaller area for the incinerator as well as for burying the residues (which are more stable). However, this method can release toxic substances into the air.

At all stages of the food chain, wastes also appear as suspended or soluble material in water. These can be treated by numerous variants of aerobic or anaerobic wastewater treatment processes. Essentially, these treatments first remove the particulate material using filtration and sedimentation and then treat the soluble material (and fine organic particles) by contacting the liquid waste with aerobic or anaerobic microorganisms. Pollution is removed from the water and "transferred" into biomass surpluses that are treated with the filtered and settled particulate matter as organic solid wastes. Variants of wastewater treatment plants exist for rich and poor countries. Their effectiveness in pollution removal is generally better in richer countries.

Gaseous nuisances are generally not treated in poorer countries as current technologies are quite expensive. These technologies rely on absorption or adsorption onto a solid or liquid substance sometimes with the aid of microorganisms. Some expensive substrates are regenerated (e.g., activated carbon) while other are cheap and can be discarded with the solid wastes (e.g., sphagnum peat biofilters).

TABLE 5: United States Environmental Protection Agency (USEPA) food recovery hierarchy

Hierarchy of action	Example of action
1-Source reduction	Reduce the amount of food waste being generated
2-Feeding people	Donation of excess food to food banks, shelters
3-Feeding animals	Provide food scraps to farmers
4-Industrial uses	Provision of fats for rendering, oil for fuel, food discards for animal feed production; anaerobic digestion with soil amendment or composting of the residues
5-Composting	Aerobic digestion of food scraps into a nutrient rich soil amendment
6-Landfill/Incineration	As last resort

Source: www.epa.gov/osw/conserve/materials/organics/food/fd-gener.htm#food-hier)

1.13 GREENING RESEARCH AND DEVELOPMENT

The development of successful food production and processing practices is anchored on good science. Research and development (R&D) are key components of the innovation continuum and in most cases is the area where the majority of funds are spent in the initial stages of process design and product development. The agricultural and agri-food sectors in particular have relied heavily on R&D to increase crop yields, decrease the need for high agricultural inputs, identify faster methods to detect pathogens, conserve foods to prevent spoilage, and identify compounds in foods with health-promoting properties, to name a few examples [6]. Activities undertaken as part of R&D efforts can have environmental impacts; thus, greening of the food production and processing sectors requires a careful look at the inputs into R&D in order to identify areas that can be targeted to reduce environmental footprints while enhancing economic benefits and speeding access to markets.

Energy and water use per square foot in many laboratory buildings are several times higher than in regular office buildings. Similarly, research and chemical laboratories use markedly larger quantities of hazardous chemicals and generate significant waste. As with food processing installations, many of these facilitates contain high numbers of containment and exhaust devices and heat-generating equipment which must be accessible continuously and, therefore, require constant maintenance and non-interrupted power supply.

Revolutions in information technology in the last two decades have provided tremendous opportunities to make R&D greener. A paradigm shift in the approaches used for experimentation has, however, been required which has made progress slow. Novel approaches worth considering include micro- and nanoscale chemistry, use of open innovation which capitalizes on networked intelligence, and adherence to concepts promoted in the waste management hierarchy presented in Fig. 5 which provides a path to minimize resource use and reduce waste. Some of these are briefly discussed below.

Microscale chemistry is an environmentally safe pollution prevention method of performing chemical processes using small quantities of chemicals without compromising the quality and standard of chemical applications in education and industry (www.microscale.org). The approach focuses on extensive reduction in the amounts of chemicals used during experimentation and further promotes the use of safe and easy-to-manipulate techniques and miniature laboratory ware allowing a decrease in chemical use by several orders of magnitude [6]. More recently, advances in nanotechnology have further expanded the possibility to miniaturise both research and development in ways that were hitherto inconceivable. Micro- and nanotechnologies are being explored in food quality, food safety, packaging, and health and nutrition research [38]. However, technological advancements in equipment design and new platforms for research, development, and "micro scale up" continue to be needed.

Another interesting trend which may offer a way to reduce resource use is multiplexing. Multiplexing is an analytical approach that allows the simultaneous detection and quantification of large number of biomolecules in small volumes of samples. Benefits of multiplexing, depending on the specific technique used, include high sensitivity and accuracy, low cost per analysis, and short time of analysis. Examples of specific applications include in immunoassays, deoxyribonucleic acid (DNA) sequence analysis, detection of toxins, allergens, pathogens, antibiotics, and among others.

For experimentation using standard approaches, the waste management hierarchy in Figure 5 can be used to decrease environmental impacts. In the first instance, employment of management strategies to minimize requirements for chemicals and hazardous materials either by avoiding use completely or reducing use is recommended. Recycling, recovering, and treatment options may be considered only if necessary followed by disposal of chemicals and hazardous materials in appropriate landfills only as the last recourse. Other approaches to be considered include the use of energy-efficient and environmentally friendly appliances and construction materials, reduced energy use for lighting, water conservation approaches, and improving building designs.

Furthermore, the use of adequate experimental designs can significantly reduce the number of tests to successfully develop a new product. In silico experimental approaches can provide insights into the viability of hypotheses without the need for experimentation which would minimize environmental impact. Additionally, developments in social media offer completely new communication and collaboration platform which opens up new models for doing business and R&D. Harnessing of ideas from the open market rather than investing millions of dollars to reinvent the proverbial R&D wheel is now considered an economically viable business model. The terms "open innovation" and "mass innovation" are increasingly being used in R&D and industry. Mass innovation refers to when a wide range of people and their different but complementary insights are brought together, to generate novel ideas by thinking outside the box [43]. Massidea.org provides an example of such a concept which is founded on series of innovation theories and is a free of charge open innovation community where people can share their ideas, discuss today's challenges as well as visions of the future, which are key factors when creating new innovations [43]. The idea to innovation continuum is very vast and requires significant financial, human, capital, energy, and time investments.

Beyond construction, chemical management, and waste minimization, greening of R&D should include efforts toward enhanced industrial collaboration, partnerships, inter-laboratory collaboration, and use of the services of specialized analytical laboratory services when appropriate [6]. To be successful in such collaborative efforts, integrity, interdependence as well as approaches for managing intellectual property (i.e., copyright, confidential information, patents, trade secrets, trademarks, designs, and plant breeder's rights) are required to ensure openness, smooth knowledge translation, and effective technology transfer.

1.14 SOCIAL PERSPECTIVES REGARDING GREEN TECHNOLOGIES

The definition of green technology or green agriculture is very fluid and often depends on the user and the context. The ultimate user of novel technologies, products, and in this instance, food is the consumer. Social perspectives of novel alternative technologies and approaches should therefore be taken into consideration in both process and product development. Consumers are confronted with an incredible amount of choice each day and their decisions are based on perceptions which may or may not be evidence-based. Choices made by consumers, nevertheless, have significant impacts on producers and processors and ultimately the economy.

Public understanding of food production and processing technologies, and their benefits and limitations will either be congruent with, or opposed to, that of scientific experts [27]. Consumers have not always lauded the intentions of the food industry due to disparities in the approaches used by the private sector. Social pressures on companies in relation to environmental issues have required that companies conform in order to avoid the consequences of non-conformity (e.g., legal penalties, public protests, and loss of market share to competitors), whereas some have responded by making fundamental changes in their practices and technologies, others avoid change through the use of public relations offensives or other avoidance tactics [27] and references therein. This partly explains the cynicism between consumers and the private sector in regard to "green claims."

Consumers are interested in the environmental impacts of agricultural and food production practices but they also require that product quality, safety, nutrition, and price not be compromised. Research shows that the acceptability of food technologies lies along a perceived risk and benefit continuum [31]. Improving benefits such as nutritional value, taste, and shelf life enhances a product's acceptability [15], whereas perceived risks (e.g., presence of toxic by-products) negatively correlates with intention to purchase, intention to use, and overall acceptability [13, 31, 47]. In general, although environmental impact of food production and processing is not the first benefit sought after by consumers in regard to food technologies, it is among their expectations [27].

The publication of Rachel Carson's *Silent Spring* revealed the dangers of chemicals used in agriculture for ecosystems and the risk they present to human health [28]. Considering the awakening provoked by what can be called the side effect of the Green Revolution, it is not surprising that the "Gene Revolution" has also aroused some suspicion. This has contributed to the growing number of consumers who turn to some sort of alternative models of food consumption such as organic agriculture, buying local, dealing directly with the producer, fair–trade, and eco-labelling (e.g., environmental and sustainable agriculture certification and evidence of social responsibility) [28].

Another major social challenge is the current trend toward using "food" for biofuels and bio-based products. At least 100 times the total global crop net primary production will be needed to meet current petroleum needs. Schenk [45] argues that there is simply not enough land to meet global fuel needs with biofuels and that we are very near the breaking point as a collective, with the real risk of not having enough food to feed everyone. A similar argument applies to bio-based plastics. Agriculture is the overwhelming source of nitrous oxide (a more potent greenhouse gas than CO_2 as previously mentioned) emissions in the United States (United States, EPA [54]. Climate friendly farming which focuses on agricultural practices which reduce greenhouse gas emissions (e.g., no till farming, mixed agriculture as opposed to monoculture) will be increasingly needed.

1.15 CURRENT AND FUTURE CHALLENGES

Continued use of current agricultural practices and technologies will not be sustainable in the long term. New business models based on finding the right balance between environment protection and economic profits are required. As shown in Fig. 6, there is a close interrelationship between food, health, economy, security and the environment. The profitability in environmental sustainability needs to become more palpable for the agriculture and agri-food sectors. To go even further, the use of agriculture not only to reduce emissions but to act as sink for other sectors through carbon sequestration must be considered. There is growing need for new technologies (e.g., through novel engineering, equipment, processing, and packaging designs) to address these emerging challenges. This is particularly of interest as various technologies will be needed to address challenges along the entire value chain. Furthermore, efforts will need to be targeted to address site-specific challenges (i.e., farm, regional, national, and international needs).

Perhaps, even more challenging will be the need for (a) technologies to accurately predict and counter expected increases in global temperatures and sea levels and attendant impacts on agriculture and capacity for processing; (b) sustainable methods to increase food yields; and (c) technologies to prevent food spoilage (especially in developing countries) where the climate is a threat to shelf life and where there is a lack of a cold chain to preserve foods. Successful diffusion and adoption of these sustainable technologies will hinge on their economic viability. Currently, there appears to be a lack of markets for environmental attributes (e.g., eco-labels with economic viability) which may need to be addressed through policy interventions (e.g., incentivization, rewards for uptake, and regulations).

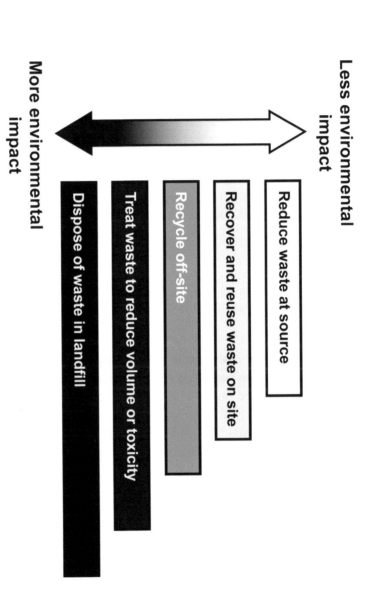

FIGURE 5: The waste management hierarchy. Source: [6]

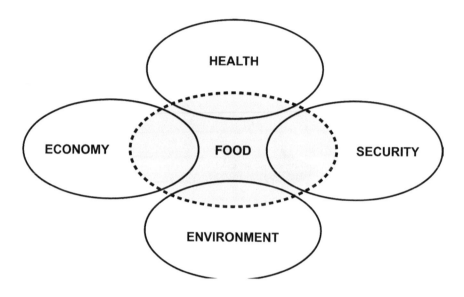

FIGURE 6: Schematic of the interrelationship between food, health, economy, security, and the environment

Another major challenge is the ongoing loss of biodiversity and its potential impacts on health and the environment. Biodiversity is being lost at an unprecedented rate as a result of current agricultural and industrial practices. Ultimately, food production and processing must be done with the objective of providing sustainable diets to all. The FAO [24] defines "sustainable diets as those diets with low environmental impacts which contribute to food and nutrition security and to healthy life for present and future generations. Such diets are protective and respectful of biodiversity and ecosystems, culturally acceptable, accessible, economically fair and affordable, nutritionally adequate, safe and healthy, while optimizing natural and human resources." Improving food systems for sustainable diets requires an interdisciplinary effort to address the problems of malnutrition, the degradation of ecosystems, and the erosion of biodiversity caused, at least in part, by modern day food systems and dietary patterns [12]. A concerted effort will thus be required to increase the diversity of foods produced in different regions of the world and to incentivize food processors to expand their use of diverse whole foods and food ingredients in food formulation and processing.

Intensive research is needed to address these challenges. As reducing cost to consumers will be important, this raises the all-important question of who must bear the cost of R&D. As indicated by the OECD, "green growth policies which place a premium on environmental protection may constrain agricultural and fisheries output, reduce global food supply and entail adjustments in the use of human, financial and natural resources in the short-term, but implications in the longer-term should be mutually reinforcing in terms of environmental sustainability, economic growth and social well-being" [41].

1.16 CONCLUSION

The relationship between food, physical and environmental health, and the economy has become increasingly evident. Finding a balance between food supply and demand in a manner that is sustainable and which ensures the long-term survival of humankind will be one of the most important challenges facing the agriculture and agri-food sectors over the next 40 years. "Increasing productivity in a sustainable manner will require according high priority to research, development, innovation, education and information in the agriculture and agri-food sectors" [41]. A variety of approaches as outlined above can be considered to reduce the impact of agricultural practices while ensuring adequate supplies of food to feed the ever growing world population. To ensure success, regional, national, and international collaborative efforts along the food chain continuum will be increasingly required. Additionally, sustainable food engineering approaches which harness the power of open innovation and which take into consideration social, environmental, economic concerns will be needed.

REFERENCES

1.   Akhtar H (2012) Reducing process-induced toxins in foods. In: Boye JI, Arcand Y (eds) Green technologies in food production and processing. Springer, New York

2. Altieri MA (1999) The ecological role of biodiversity in agroecosystems. Agric Ecosyst Environ 74:19–31
3. Arcand Y, Maxime D, Zareifard R (2012) LCA of processed food. In: Boye JI, Arcand Y (eds) Green technologies in food production and processing. Springer, New York
4. Bloom J (2010) American wasteland: how America throws away nearly half of its food (and what we can do about it). Da Capo Press, Boston, MA
5. Bouwman AF, Boumans LJM, Batjes NH (2002) Modeling global annual N2O and NO emissions from fertilized fields. Global Biogeochem Cycles 16(4):1080
6. Boye JI, Maltais A, Bittner S, Arcand Y (2012) Greening of research and development. In: Boye JI, Arcand Y (eds) Green technologies in food production and processing. Springer, New York
7. Boye JI, Arcand Y (2012) Green technologies in food production and processing. Springer, New York
8. Brentrup F, Küsters J, Lammel J, Kuhlmann H (2000) Methods to estimate on-field nitrogen emissions from crop production as input to LCA studies in the agricultural sector. Int J Life Cycle Assess 5:349–357
9. Brentrup F, Küsters J, Lammel J, Kuhlmann H (2002) Life cycle impact assessment of land use based on the Hemeroby concept. Int J Life Cycle Assess 7:339–348
10. Brentrup F (2012) LCA of crop production. In: Boye JI, Arcand Y (eds) Green technologies in food production and processing. Springer, New York.
11. Bruntland Commission (1987) Our common future: report of the world commission on environment and development, published as annex to general assembly document A/42/427, development and international cooperation. Oxford University Press, New York
12. Burlingame B, Charrondiere UR, Dernini S, Stadlmayr B, Mondovì S (2012) Food biodiversity and sustainable diets: implications of applications for food production and processing. In: Boye JI, Arcand Y (eds) Green technologies in food production and processing. Springer, New York
13. Cardello AV, Schutz HG, Lesher LL (2007) Consumer perceptions of foods processed by innovative and emerging technologies. A conjoint analytic study. Innov Food Sci Emerg Technol 8(1):73–83
14. Conant RT, Paustian K (2002) Potential soil carbon sequestration in overgrazed grassland ecosystems. Global Biogeochem Cycles 16(4):1143–1152
15. Delgado-Gutierrez C, Bruhn CM (2008) Health professionals' attitudes and educational needs regarding new food processing technologies. J Food Sci Educ 7(4):78–83
16. Dyer JA (1989) A canadian approach to drought monitoring for famine relief in Africa. Water Int 14(4):198–205
17. European Commission—Joint Research Centre—Institute for Environment and Sustainability (2010) International reference life cycle data system (ILCD) handbook—general guide for life cycle assessment—detailed guidance. First edition March 2010. EUR 24708 EN. Publications Office of the European Union; Luxembourg
18. Fearnside PM (1980) The effect of cattle pasture on soil fertility in the Brazilian Amazon: consequence for beef production fertility. Trop Ecol 21(1):125–137

19. Fearnside PM (1990) Deforestation in Brazilian Amazonia. In: Woodwell GM (ed) The earth in transition: patterns and processes of biotic impoverishment. Cambridge University Press, New York

20. Food and Agriculture Organization of the United Nations (FAO) (2003) World agriculture: towards 2015/2030, an FAO perspective. Earthscan Publications Ltd, London

21. Food and Agriculture Organization of the United Nations (FAO) (2006) World agriculture: towards 2030/2050, an FAO perspective. Interim report

22. Food and Agriculture Organization of the United Nations (FAO) (2009a) The state of food in agriculture: livestock in the balance. Electronic Publishing Policy and Support Branch-Communication Division, FAO, Rome

23. Food and Agriculture Organization of the United Nations (FAO) (2010) FAOSTAT statistical database. Rome. Accessed Mar 2010

24. FAO (2010) International scientific symposium 'biodiversity and sustainable diets', final document. Available at http://www.fao.org/ag/humannutrition/25915-0e8d-8dc364ee46865d5841c48976e9980.pdf)

25. Ferguson A (2012) Population matters for a sustainable future. OPT J 12(2):4–6

26. Gagnon N (2012) Introduction to the global agri-food system. In: Boye JI, Arcand Y (eds) Green technologies in food production and processing. Springer, New York

27. Gauthier EG (2012) Green food processing technologies: factors affecting consumers' acceptance. In: Boye JI, Arcand Y (eds) Green technologies in food production and processing. Springer, New York

28. Gendron C, Audet R (2012) Key drivers of the food chain. In: Boye JI, Arcand Y (eds) Green technologies in food production and processing. Springer, New York

29. Grabowski S, Boye JI (2012) Green technologies in food dehydration. In: Boye JI, Arcand Y (eds) Green technologies in food production and processing. Springer, New York

30. Grabowski S, Marcotte M (2003) Pretreatment efficiency in osmotic dehydration of cranberries. In: Chanes JW, Velez-Ruiz JF, Barbosa-Canovas GV (eds) Transport phenomena in food processing. CRC Press, Boca Raton

31. Henson S, Annou M, Cranfield J, Ryks J (2008) Understanding consumer attitudes toward food technologies in Canada. Risk Anal 28(6):1601–1617

32. Hermansen JE, Nguyen TLT (2012) LCA and the agri-food chain. In: Boye JI, Arcand Y (eds) Green technologies in food production and processing. Springer, New York

33. Hoekstra AY, Chapagain AK (2007) Water footprints of nations: water use by people as a function of their consumption pattern. Water Resour Manag 21:35–48

34. Hunt R, Franklin W (1996) LCA—how it came about. Int J Life Cycle Assess 1(1):4–7

35. LeJeune JT (2003) E. coli O157 in dairy cattle: questions and answers. Ohio State University Extension Fact Sheet. VME-16-03. Food Animal Health Research Program, Wooster, Ohio. http://ohioline.osu.edu/vme-fact/0016.html. Accessed 15 Jun 2010

36. Mena C, Stevens G (2010) Delivering performance in food supply chains: an introduction. In: Mena C, Stevens G (eds) Delivering performance in food supply chains, 1st edn. Woodhead Publishing, London

37. Menzi H, Oenema O, Burton C, Shipin O, Gerber P, Robinson T, Franceschini G (2010) Impacts of intensive livestock production and manure management on ecosystems, chapter 9. In: Steinfeld H, Mooney AH, Schneider F, Neuville EL (eds)

Livestock in a changing landscape, vol. 1: Drivers, consequences and responses. Island Press, Washington, DC, pp 139–163

38. Nazzaro F, Fratianni F, Coppola R (2012) Nano and micro analyses. In: Boye JI, Arcand Y (eds) Green technologies in food production and processing. Springer, New York

39. Nellemann C, MacDevette M, Manders T, Eickhout B, Svihus B, Prins AG, Kaltenborn, BP (2009) The environmental food crisis—the environment's role in averting future food crises: a UNEP rapid response assessment. United Nations Environment Programme (UNEP)/GRID-Arendal, Birkeland Trykkeri AS, Norway

40. Ngadi MO, Latheef MB, Kassama L (2012) Emerging technologies for microbial control in food processing. In: Boye JI, Arcand Y (eds) Green technologies in food production and processing. Springer, New York

41. OECD (2011) A green growth strategy for food and agriculture: Preliminary report. OECD 2011. www.oecd.org

42. Pirog R, Van Pelt T, Ensayan K, Cook E (2001) Food, fuel and freeways: An Iowa perspective on how far food travels, fuel usage, and greenhouse gas emissions. Leopold Center for Sustainable Agriculture, Ames

43. Santonen T (2012) Massidea.org—a greener way to innovate. In: Boye JI, Arcand Y (eds) Green technologies in food production and processing. Springer, New York

44. Schader C, Stolze M, Gattinger A (2012) Comparison of traditional vs alternative (organic) farming. In: Boye JI, Arcand Y (eds) Green technologies in food production and processing. Springer, New York

45. Schenk R (2012) The fallacy of biobased materials. In: Boye JI, Arcand Y (eds) Green technologies in food production and processing. Springer, New York

46. Selke SEM (2012) Green packaging. In: Boye JI, Arcand Y (eds) Green technologies in food production and processing. Springer, New York

47. Siegrist M, Cousin ME, Kastenholz H, Wiek A (2007) Public acceptance of nanotechnology foods and food packaging. The influence of affect and trust. Appetite 49:459–466

48. Simpson BK, Rui X, XiuJie J (2012) Enzyme-assisted food processing. In: Boye JI, Arcand Y (eds) Green technologies in food production and processing. Springer, New York

49. Smith P, Martino D, Cai Z, Gwary D, Janzen H, Kumar P, McCarl B, Ogle S, O'Mara F, Rice C, Scholes B, Sirotenko O (2007) Agriculture. In: Metz B, Davidson OR, Bosch PR, Dave R, Meyer LA (eds) Climate change 2007: mitigation. Contribution of working group III to the fourth assessment report of the intergovernmental panel on climate change. Cambridge University Press, Cambridge

50. Seinfeld H, Gerber P, Wassenaar T, Castel V, Rosales M, de Haan C (2006) Livestock's long shadow: environmental issues and options. FAO, Rome

51. Stevens GC (1989) Integrating the supply chain. Intern J Phys Distrib Mater Manag 19(8):3–8

52. Stolze M, Piorr A, Häring AM, Dabbert S (2000) Environmental impacts of organic farming in Europe. Organic farming in Europe: economics and policy, vol. 6. Universität Stuttgart-Hohenheim, Stuttgart-Hohenheim

53. UNECE/EMEP (2007) EMEP/CORINAIR emission inventory guidebook—2007. Technical report No 16/2007. EEA (European Environment Agency), Copenhagen, Denmark

54. United States. Environmental Protection Agency (2011) Draft inventory of the U.S. greenhouse gas emissions and sink: 1990–2009. http://www.epa.gov/climatechange/emissions/downloads11/US-GHG-Inventory-2011-Complete_Report.pdf

55. Vayssières J, Rufino MC (2012) Management of agricultural inputs, waste and farm outputs: present and future best management practices. In: Boye JI, Arcand Y (eds) Green technologies in food production and processing. Springer, New York

56. Vergé XPC, Worth DE, Desjardins RL, McConkey BG, Dyer JA (2012) LCA of animal production. In: Boye JI, Arcand Y (eds) Green technologies in food production and processing. Springer, New York

57. Wakeland W, Cholette S, Venkat K (2012) Transportation issues and reducing the carbon footprint. In: Boye JI, Arcand Y (eds) Green technologies in food production and processing. Springer, New York

58. Weber CL, Matthews HS (2008) Food-miles and the relative climate impacts of food choices in the United States. Environ Sci Technol 42(10):3508–3513

59. World Health Organization (WHO)/Food and Agriculture Organization of the United Nations (FAO) (2003) Diet, nutrition and the prevention of chronic diseases. WHO/FAO Expert Consultation, Geneva, Switzerland. (WHO Technical Report Series;916)

60. Zervas G (1998) Quantifying and optimizing grazing regimes in Greek mountain systems. J Appl Ecol 35:983–986

CHAPTER 2

Design of Sustainable Supply Chains for the Agrifood Sector: A Holistic Research Framework

E. IAKOVOU, D. VLACHOS, CH. ACHILLAS, AND F. ANASTASIADIS

2.1 INTRODUCTION

The agrifood industry is a sector of key economic and political importance. It is one of the most regulated and protected sectors in the EU, with major implications for sustainability such as the fulfillment of human needs, the support of employment and economic growth, and its impact on the natural environment. According to the European Commission, more than 17 million operators and 32 million individuals are involved across the food chain (European Communities, 2008). Moreover, the food and drink sector contributes to 20%-30% of all environmental impacts in EU (Bakas, 2010). Growing environmental, social and ethical concerns, and increased awareness of the effects of food production and consumption on

Design of Sustainable Supply Chains for the Agrifood Sector: A Holistic Research Framework. © *Iakovou E, Vlachos D, Achillas Ch, and Anastasiadis F.* Agricultural Engineering International: CIGR Journal **Special Issue 2014** *(2014). Licensed under Creative Commons Attribution License, http://creativecommons.org/licenses/by/3.0/.*

the natural environment have led to increased pressure by consumer organizations, environmental advocacy groups, policy-makers, and several
consumer groups on agrifood companies to deal with social and environmental issues related to their supply chains within product lifecycles, from
'farm to fork' (Courville, 2003; Weatherell and Allinson, 2003; Ilbery and
Maye, 2005; Maloni and Brown 2006; Vachon and Klassen, 2006; Welford and Frost, 2006; Matos and Hall, 2007).

Sustainability of supply chain management had gained a lot of academic and business interest during the last years (Seuring, 2012). Seuring and Müller (2008) presented a comprehensive literature review with
191 relevant papers and outline the major lines of research in the field.
Moreover, in the work of Gupta and Palsule-Desai (2011), the existing
literature is taxinomized under four broad categories, namely strategic
considerations, decisions at functional interfaces, regulation/government
policies, and decision support tools. The aim is to provide managers and
practitioners with the most important issues in sustainable supply chain
management decision-making. Similarly, Seuring (2012) reviewed papers
on sustainable supply chains which apply quantitative models.

The aim of the proposed framework is the optimization of the agrifood
supply chain design, planning and operations through the implementation
of appropriate green supply chain management and logistics principles.
The research objective is the minimization of the environmental burden
and the maximization of supply chain sustainability of the selected product categories.

The rest of the paper is organized as follows. In Section 2, we present the
emergence of green supply chain management as a key corporate strategic
priority and a center of profitability, while we further focus on its importance
on the agrifood sector. The proposed holistic methodological framework encompassing six thematic areas is analyzed in Section 3. Finally, we sum-up
with conclusions and future research directions in Section 4.

2.2 EMERGENCE OF GREEN SUPPLY CHAINS

Although the importance of the research focal issue, that of reducing and
controlling the environmental footprint of agrifood supply chains, is now

recognized even from the laymen, herein we further document its value by providing a few characteristic relevant data and by summarizing the results of recent research efforts.

Today, societal stakeholders demand corporate responsibility to transcend product quality and rather extend to areas of labor standards, health and safety, environmental sustainability, non-financial accounting and reporting, procurement, supplier relations, product lifecycles and environmental practices (Bakker and Nijhof, 2002; Waddock and Bodwell, 2004; Teuscher et al., 2006). Sustainable supply chain management expands the concept of sustainability from a company to the supply chain level (Carter and Rogers, 2008) by providing companies with tools for improving their own and the sector's competitiveness, sustainability and responsibility towards stakeholder expectations (Fritz and Schiefer, 2008). Principles of accountability, transparency and stakeholder engagement were highly relevant to sustainable supply chain management (Waddock and Bodwell, 2004; Teuscher, 2006; Carter and Rogers, 2008).

More specifically, in response to pressures for transparency and accountability, agrifood companies need to measure, benchmark, and report environmental sustainability performance of their supply chains; whilst on the other side, policy-makers need to measure the sectorial performance within the supply chain context for effective target setting and decision-making interventions.

Furthermore, in order to unleash value, it is important to exploit the potential of utilizing agrifood waste and the associated by-product biomasses for energy recovery and nutrient recycling, to mitigate climate change and eutrophication which are currently unexploited (Kahiluoto et al., 2011). To that end, biomass emerges as a promising option, mainly due to its potential worldwide availability, its conversion efficiency and its ability to be produced and consumed on a CO_2-neutral basis. Biomass is a versatile energy source, generating not only electricity but also heat, while it can be further used to produce biofuels (Verigna, 2006). Iakovou et al. (2010) provided a critical synthesis of the state-of-the-art literature on waste biomass supply chain management. Agrifood biomass is usually free of toxic contaminants and is determined spatially and temporally by the respective local/regional profile of the pertinent activities. It is well documented that 31% of the greenhouse-gas emissions and more than 50% of eutrophication are related to food chains, thus highlighting the need to intervene in the agrifood supply chain to ameliorate its impact on the environment (CEC, 2006). In order to promote "green" agrifood supply chains (GAF-SCs) and elaborate agrifood biomass operations on large scale, the application of appropriately designed innovative policies and systems is necessary (van der Vorst et al., 2009; Negro and Smits, 2007).

Additionally, the recent post-2009 recession period has further underlined the need to turn the business focus, across the world, not only to profitability, but to sustainability as well. Today, one of the key priorities in corporate strategic design for an organization is to emerge as socially responsible and sustainable through environment protection. Companies are structuring their sustainability reports disclosing their strategy to address the growing concerns of environmental degradation and global warming. Today, 80% of the global Fortune 250 companies release their annual sustainability report, up from 37% in 2005 (Singh, 2010). As a focal part of sustainability initiatives, green supply chain management has emerged as a key strategy that can provide competitive advantage with substantial gains for the company's bottom line. In designing green supply chains, the intent is to adopt comprehensively and across business boundaries, best practices right from product conception to the end-of-life recycling stage. Under this context, green initiatives relate to tangible and intangible corporate benefits. Sustainability reports of many companies indicate that the greening of their supply chains has helped them to reduce their operating cost with increased sustainability of their business.

The result of a recent survey conducted by McKinsey documents that green supply chain management is one of the top two strategic priorities for global corporations (McKinsey, 2011). The benefits of going green are substantial. A green supply chain can not only reduce an organization's carbon footprint but also lead to reduced costs, improved reputation with customers, investors and other stakeholders, thus leading to a competitive edge in the market and therefore increased profitability.

The importance of linking research to sustainable development is strongly acknowledged, and the framework for doing so at the EU level has been set up reciprocally in the EU renewed Sustainable Development Strategy and in the Seventh Framework Programme. This is further reaffirmed in most recent EU R&D policy documents; the Communication on

"A Strategic European Framework for International Science and Technology Cooperation" and the Communication on "Toward Joint Programming in Research: Working together to tackle common challenges more effectively". Furthermore, the ERA vision 2020 (within the Ljubljana process) calls for the European Research Area to focus on society's needs and ambitions towards sustainable development. The three Key Thrusts identified by ETP Food for Life Strategic Research Agenda 2007-2020 (SRA) meet all of the criteria required to stimulate innovation, creating new markets and meet important social and environmental goals. These Thrusts are:

1. Improving health, well-being and longevity;
2. Building consumer trust in the food chain;
3. Supporting sustainable and ethical production.

According to the third Key Thrust, food chains should operate in a manner that exploits and optimizes the synergies among environmental protection, social fairness and economic growth. This will further ensure that the consumers' needs for transparency and for affordable food of high quality and diversity are fully met. Progress in this area is expected to have important benefits for the industry in terms of reduced uses of resources, increased efficiency and improved governance. In July 2008, the European Commission adopted action plans for the Sustainable Consumption and Production (SCP) and a Sustainable Industrial Policy (SIP). The plans followed a 2005 Commission communication on a thematic strategy for the sustainable use of natural resources, which calls for sectorial initiatives to be launched together with economic operators. A European Retail Forum and Retailers' Environmental Action Programme (REAP) were launched in 2008 to promote voluntary action to reduce the environmental footprint of the retail sector and its supply chain, to promote more sustainable products, and to support consumers buying "green". In May 2009, the EU sustainable food chain roundtable was launched seeking to develop a methodology for assessing the environmental footprint of individual foods and drinks by 2011. The roundtable brought together farmers and suppliers, food and drink producers, packaging firms and consumer organizations to develop environmental assessment methodologies for products and means for effective consumer communication, and to report on improvements.

An overview of emerging global trends, policy developments, challenges and prospects for European agri-futures, point to the need for new strategic frameworks for the planning and delivery of research. Such frameworks should address the following challenges:

1. Sustainability: facing climate change in the knowledge-based bio-society;
2. Security: safeguarding European food, rural, energy, biodiversity and agri-futures;
3. Knowledge: user-oriented knowledge development and exchange strategies;
4. Competitiveness: positioning Europe in agrifood and other agricultural lead markets;
5. Policy and institutional: facing policy-makers in synchronizing multi-level policies.

Addressing these challenges can shift the European agrifood sector to the knowledge-based bio-economy, while satisfying the need for the sector (and food retailers) to remain globally competitive while addressing climate change and sustainable development concerns, such as the maintenance of biodiversity and prevention of landscape damage. Addressing these multi-faceted sustainable development challenges facing the agrifood sector in Europe and worldwide, will require a major overhaul in the current agriculture research system. Recent foresight work under the aegis of Europe's Standing Committee for Agricultural Research (SCAR) has highlighted that in the emerging global scenario for European agriculture, research content needs to extend to address a diverse and often inter-related set of issues relating to sustainable development, including food safety/security, environmental sustainability, biodiversity, bio-safety and bio-security, animal welfare, ethical foods, fair trade and the future viability of rural regions. These issues cannot simply be added to the research agenda; addressing them comprehensively and holistically in agriculture research requires new methods of organizing research, in terms of priority-setting, research evaluation and selection criteria, and in bringing together new configurations of research teams, as well as managing closer interactions with the user communities and the general public in order to ensure that relevant information and knowledge is produced and the results are properly disseminated.

Although sustainability and environmental impact assessments have traditionally focused on agriculture (McNeeley and Scherr, 2003; Filson, 2004), researchers and policy-makers have recently made attempts to develop more systemic approaches by incorporating stages of food processing, food retailing and specifically transportation in the assessment frameworks of food supply chains (Bakker and Nijhof, 2002; Heller and Keoleian, 2003; Green and Foster, 2005). Various approaches have been developed to measure sustainability of the food supply chains that identify effects at regional, industrial, and firm levels. Some specific sustainability assessment frameworks developed for the food sector include: farm economic costing (Pretty et al., 2005); lifecycle approach to sustainability impacts (Heller and Keoleian, 2003; Blengini and Busto, 2009; Roy et al., 2009); food miles (Coley et al., 2009; Kemp et al., 2010); energy accounting in product lifecycle (Carlsson-Kanayama et al., 2003); mass balance of food sectors (Risku-Norjaa and Mäenpääb, 2007; Lopez et al., 2008; Ortiz, 2008); ecological footprint (Gerbens-Leenes et al., 2002; Collins and Fairchild, 2007; Burton, 2009; Ridoutt et al., 2010; Mena et al., 2011); and farm sustainability indicators (Fernandes and Woodhouse, 2008; Meul et al., 2009; Nickell et al., 2009; Gómez-Limón and Sanchez-Fernandez, 2010; Rodrigues et al., 2010).

Finally, there has been an emergent set of research efforts related to benchmarking and performance measurement. However, most of this research is oriented towards the improvement of individual firms or processes rather than the analysis of entire supply chains (McNeeley and Scherr, 2003; Filson, 2004). Few efforts have been made to measure supply chain performance, while the focus has primarily been on efficiency and other economic-related performance, whereas in the current research framework there is a strong emphasis on environmental performance. Thus, there is a need to capture environmental performance throughout the entire supply chain. The enhancement of such measurements by incorporating stakeholder aspects and additional environmental dimensions is rare or does not exist at all (Bakker and Nijhof, 2002).

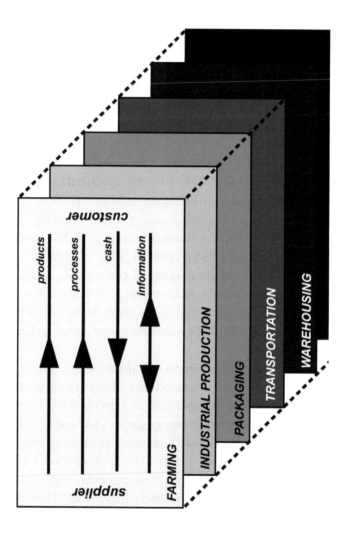

FIGURE 1: Supply chain management echelons.

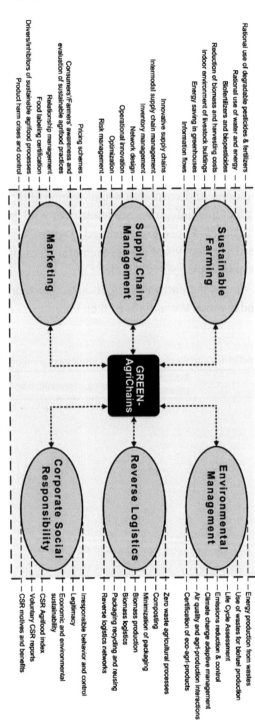

FIGURE 2: Conceptual framework.

2.3 HOLISTIC METHODOLOGICAL FRAMEWORK

Figure 1 exhibits conceptually the main echelons encountered in agrifood supply chains. A comprehensive framework that tackles holistically and interdisciplinary all aspects of green supply chain management in the agrifood sector should be spanning across: sustainable farming, reverse logistics (waste management and packaging reuse), green procurement and sourcing, transportation, energy consumption efficiency, green marketing, green accounting, and corporate social responsibility (CSR). To that effect, six distinct thematic areas are identified, with each of them having a number of issues that need to be tackled (Figure 2).

The interdependencies of the six thematic areas and their impact on the six supply chain management echelons are captured in Table 1.

Each of the six thematic areas is discussed further below, aiming to reveal the need towards the initiative taken to optimize the production chain.

TABLE 1: Benefits for supply chains from the implementation of green practices

	Farming	Industrial Production	Packaging	Transportation	Warehousing	Distribution
Supply Chain Management		x	x	x	x	
Sustainable Farming	x					
Reverse Logistics	x	x	x	x	x	x
Marketing		x	x			
Environmental Management	x	x	x	x	x	x
Corporate Social Responsibility	x	x	x	x	x	x

2.3.1 SUPPLY CHAIN MANAGEMENT

Focus needs to be given on sustainability improvement of supply chain and logistics operations in the agrifood industry, including research in supporting information systems and reducing the energy and pollution from transportation. Although most of the problems are sector-independent, there are few unique characteristics of the agrifood industry that differentiate traditional approaches. Such characteristics include the perishability of most agrifood products that highlight the importance of timely delivery as well as the need for developing "cold" supply chains and the requirement for product traceability along the supply chain, which is closely related to the visibility of supply chain. Indicatively, Sarkis et al. (2011) and Seuring and Müller (2008) presented a comprehensive review of issues that need to be tackled within this thematic area.

2.3.2 SUSTAINABLE FARMING

Agriculture is one of the most important contributors to today's most serious environmental problems. The use of chemicals pesticides for the weed and the pest control, the use of artificial fertilizers, the improper management of animal wastes and other wastes produced from biomass production and the use of high levels of water for irrigation, led to the degradation of the rural environment. Moreover, agriculture consumes considerable amounts of energy, either directly for operating machinery and equipment on the farm, as well as for heating of agricultural buildings (greenhouses, livestock buildings, etc.) or indirectly for the production of fertilizers and pesticides used in the crops.

Reduction of the energy use in agriculture in a sustainable manner is attained by the energy production (methane and biohydrogen) through the anaerobic degradation of the organic wastes, by the use of energy saving systems in agricultural buildings and of innovative systems for harvest and tillage. The bio-fertilizers produced after the fermentation of the animal wastes can be used instead of artificial fertilizers, the high amounts of wastewater after treatment can be used for irrigation purposes, and the use of an integrated farming system including crop rotation could minimize the use of chemical pesticides for weed and pest control. The adoption of these practices can play an important role towards attaining sustainable agriculture. Indicatively, in the work of Acs et al. (2005), the technical, economic, and environmental aspects of organic farming are thoroughly assessed.

2.3.3 REVERSE LOGISTICS

Reverse logistics presents a critical area towards green supply chains for the agrifood sector. A special focus needs to be placed on reusing agrifood containers and recycling packaging materials or re-designing packaging to use less material. Additionally, all the operations linked to the reuse of products and materials in the agrifood supply chain, for example, the logistics activities of collecting and processing of products/ materials and used pieces, should be examined towards the direction of reassuring their sustainable restoration. Indicatively, critical issues within reverse logistics are investigated through content analysis of the published literature in the work of Pokharel and Mutha (2009).

2.3.4 MARKETING

The main focus regarding this area is on market performance, pricing policies and customers' satisfaction in the agrifood supply process. Goals should include inter alia pricing, relationship management (covering numerous stakeholders such as producers, suppliers and consumers). Auspiciously, Johns and Pine (2002) reviewed the literature relating to consumer studies in food industry. Specific issues that need to be tackled are:

1. Pricing scenarios based on the food characteristics (organic products presented as premium products serving niche markets) and the methods adopted for their production (e-labeling, soil fertilized with by-products, recycled water, etc.).
2. Consumers' attitudes and behaviors towards products that result from sustainable ways of production, (i.e., products grown with renewable energy, for instance, recycled water, photovoltaic, biomass used as fertilizer, etc.).
3. Consumers' attitudes towards eco-labeling, food safety assurance, agrifood standards, and third-party certification.
4. Drivers and inhibitors of sustainable agrifood productions (elements such as ethics, social values, sustainability attitudes, trust, social desirability, image management constructs are considered).

5. Consumers' knowledge of organic products selling points in order to increase their selling power/efficiency.
6. Consumers' knowledge and attitudes towards agriculture entities' Corporate Social Responsibility (CSR) and Corporate Social Irresponsibility activities (CSI).
7. Whether CSR serves as a protection measurement against product harm crises (such as suppliers' and consumers' outcries and boycotts).
8. Consumers'/farmers' willingness to consume/ produce food grown with renewable energy sources (for instance recycled water, photovoltaic, biomass used as fertilizer etc.).

2.3.5 ENVIRONMENTAL MANAGEMENT

An area of great concern is associated with biodiversity, soil quality and water habitats as well as the emissions due to production and logistics operations. Environmental management of supply chains is assessed in numerous studies during the past years. Indicatively, Hassini et al. (2012) reviewed the literature on sustainable supply chains during the last decade, while Walker et al. (2008) studied the critical factors towards the implementation of "green" supply chain management initiatives. Within the proposed framework, the following issues need to be addressed:

1. Rational use of pesticides and fertilizers.
2. Rational water and energy use: consumption and nature of raw materials (including water) used in agrifood production and their energy efficiency, best irrigation practices, water planning, crop management plan.
3. Life Cycle Analysis: assessment of agrifood environmental burden throughout products' life cycle (from cradle to grave), applications of the LCA methodology to food product systems and to food consumption patterns, support of information sharing and exchange of experience regarding environmental conscious decision-making in the agrifood chain, provision of background for the sustainability of the agrifood sector.

4. Emissions reduction and control: best available techniques, green-house gas emissions mitigation strategies, economic and technical viability of upgrading existing installations, use of low-waste technology/less hazardous substances, comparable processes/facilities/methods, technological advances and late changes.

5. Climate change adaptive management: impacts of climate changes on different ecosystems, consequences to agricultural production, changes in the seasonal and annual patterns of agricultural production, extreme weather events and disaster management, adaptation measures towards climate change.

6. Interactions between air quality and agri-production: crop damages from air pollution, forecasting of agricultural production, quality of food production.

7. Certification of eco-agrifood: eco-labeling, tracing of food and feed, food safety assurance, agrifood standards, third-party certification.

2.3.6 CORPORATE SOCIAL RESPONSIBILITY (CSR)

The mitigation of irresponsible behavior, opportunities for corporations' legitimacy, commitment of agriculture business to economic and environmental sustainability (harmonious use of environmental and human resources, i.e., use of local communities, work equality, work opportunities to both genders, respect of minorities etc.) should be thoroughly explored. Specific issues that need to be tackled are:

1. Mitigation of resources waste, use of alternative eco-friendly power, equal opportunities (work and supply), respect of local communities (e.g., local small farmers), promotion of environmentally friendly-farming methods.

2. Use of CSR activities to promote corporate actions and strategies but not in the expense of society's interests and well-being (pollution, considering resource scarcity, i.e., use of recycling water).

3. Use of CSR activities for corporations' legitimization. For instance, large corporations could be particularly benefited, while small and medium sized enterprises could also use them as promotional tools.

4. Establish CSR agriculture Index tackling the following issues: (a) beneficial product and services, (b) pollution prevention, (c) recycling (of resources and byproducts), (d) clean energy, and (e) management systems which target social equality.

5. Production of voluntary CSR reports. A CSR publication provides accountability over and above legal obligations while competition pressures are alleviated.

6. Relationships among CSR activities, financial performance, sales increase and consumers' satisfaction/loyalty.

7. Comparing and contrasting agriculture entities' CSR and Corporate Social Irresponsibility.

8. Criteria for the detection of cases where CSR activities are intended to mask Corporate Social Irresponsibility.

9. Agriculture CSR resulting benefits (achievement of relationship management with customers, suppliers, sellers etc.).

10. Corporate Social Irresponsibility actions and their potential outcomes (such as boycotts, effects on brand image, pricing policies, and advertising etc.).

11. Adoption of CSR activities as protection measurements against product (harms) crises (such as suppliers' and consumers' outcries and boycotts).

Indicatively, Kong (2012) and Cuganesan et al. (2010) analytically examined Corporate Social Responsibility issues within the agri-food industry.

2.4 CONCLUSIONS

The proposed framework for the optimized design of green supply chains for the agrifood sector is expected to foster sustainable regional economic and social development in two major axes, namely rural development and agriculture sector. Taking into account that over 60% of the population of the in the EU-27 resides in rural areas, which cover 90% of the EU territory, the rural development is a vitally important policy area. Farming and forestry remain crucial sectors for the land use and the management of natural resources in the EU's rural areas. These sectors can be, also, considered as well as a platform for economic diversification in rural communities. The strengthening of rural development policy has, therefore, become an overall EU priority. The proposed framework is focused on the development of state-of-the-art supply chain management methodologies for increasing farmers' income through the optimization of the farming operations and through the reduction of the operational cost in the farm. Biomass or biofuel production can also have a positive impact on agricultural employment and rural development, particularly when conversion facilities are of smaller-scale and are, also, located near crop sources in rural districts. Finally, new crops can, also, be introduced as economically profitable alternatives to declining crops (i.e. cotton), according to the European CAP (Common Agricultural Policy).

In respect to sustainable development, the proposed framework needs to focus on the development of green operations that will lead to new environmentally benign supply chain design and operations replacing less sustainable practices. Moreover, the application of such a comprehensive framework could result into major reduction of CO_2 emissions, helping the EU to achieve at least a 20% reduction of greenhouse gases by 2020 compared to 1990 levels and an objective for a 30% reduction by 2020. This may be achieved both by the additional production of others biofuels from wastes, as well as the introduction of a novel intelligent logistics network, in order to reduce the harvest and transportation energy input. Last but not least, the expansion of the biomass feedstock available for biofuel production can provide adequate support towards avoidance of food/fuel competition for land use. The impact of the proposed framework on the Environment and Sustainable Development is thus in accordance with a number of EU policies, such as Environmental Technologies Action Plan, Common Agricultural Policy, Climate action and renewable energy package and the EU Sustainable Development Strategy.

REFERENCES

1.   Acs, S., P. B. M. Berentsen, and R. B. M. Huirne. 2005. Modelling conventional and organic farming: a literature review. NJAS - Wageningen Journal of Life Sciences, 53(1): 1-18.

2. Bakas, I. 2012. Food and Greenhouse Gas (GHG) Emissions [Online]. Available: http://www.scp-knowledge.eu/sites/default/files/KU_Food_GHG_emissions. pdf (accessed August 22, 2012).

3. Bakker, F. d., and A. Nijhof. 2002. Responsible chain management: a capability assessment framework. Business Strategy and the Environment, 11(1):63-75.

4. Blengini, G. A., and M. Busto. 2009. The life cycle of rice: LCA of alternative agrifood chain management systems in Vercelli (Italy). Journal of Environmental Management, 90 (3):1512-1522.

5. Burton, C. H. 2009. Reconciling the new demands for food protection with environmental needs in the management of livestock wastes. Bioresource Technology, 100(22): 5399-5405.

6. Carlsson-Kanayama, A., M. P. Ekstrom, and H. Shanahan. 2003. Food and life cycle energy inputs: Consequences of diet and ways to increase efficiency. Ecological Economics, 44(2-3): 293-307.

7. Carter, C. R. and Rogers, D. S. 2008. A framework of sustainable supply chain management: moving towards new theory. International Journal of Physical Distribution & Logistics Management, 38(5): 360-387.

8. CEC. 2006. Environmental Impact of Products of Products (EIPRO). Analysis of consumption of the EU-25. Technical Report EUR 22284.

9. Coley, D., M. Howard, and M. Winter. 2009. Local food, food miles and carbon emissions: A comparison of farm shop and mass distribution approaches. Food policy, 34(2): 150-155.

10. Collins, A., and R. Fairchild. 2007. Sustainable food consumption at a sub-national level: an ecological footprint, nutritional and economic analysis. Journal of Environmental Policy and Planning, 9(1): 5-30.

11. Courville, S. 2003. Use of indicators to compare supply chains in the coffee industry. Greener management international, 43(13): 94-105.

12. Cuganesan, S., J. Guthrie, and L. Ward. 2010. Examining CSR disclosure strategies within the Australian food and beverage industry. Accounting Forum, 34(3-4): 169-183.

13. European Communities. 2008. Food: from farm to fork statistics. European pocketbooks. Luxembourg.

14. Fernandes, L. A. O., and P. J. Woodhouse. 2008. Family farm sustainability in southern Brazil: An application of agri-environmental indicators. Ecological Economics, 66(2-3): 243-257.

15. Filson, G. C. 2004. Intensive Agriculture and Sustainability: A Farming Systems Analysis. Vancouver, UBC Press.

16. Fritz, M., and G. Schiefer. 2008. Food chain management for sustainable food system development: a European research agenda. Agribusiness, 24(4):440-452.

17. Gerbens-Leenes, P. W., S. Nonhebel, and W. P. M. F. Ivens. 2002. A method to determine land requirements relating to food consumption patterns. Agriculture, Ecosystems and Environment, 90(1): 47-58.

18. Gómez-Limón, J. A., and G. Sanchez-Fernandez. 2010. Empirical evaluation of agricultural sustainability using composite indicators. Ecological Economics, 69(5): 1062-1075.

19. Green, K. and Foster, C. 2005. Give peas a chance: transformations in food consumption and production systems. Technological Forecasting and Social Change, 72(6): 663- 679.

20. Gupta, S. and Palsule-Desai, O. 2011. Sustainable supply chain management: Review and research opportunities. IIMB Management Review, 23(4): 234-245.

21. Hassini, E., Surti, C., and C. Searcy. 2012. A literature review and a case study of sustainable supply chains with a focus on metrics. International Journal of Production Economics, 140(1): 69-82.

22. Heller, M. C., and G. A. Keoleian. 2003. Assessing the sustainability of the US food system: a life cycle perspective. Agricultural Systems, 76(3): 1007-1041.

23. Ilbery, B., and D. Maye. 2005. Food supply chains and sustainability: evidence from specialist food producers in the Scottish/English borders . Land Use Policy, 22(4): 331-344.

24. Johns, N., and R. Pine. 2002. Consumer behaviour in the food service industry: a review. International Journal of Hospitality Management, 21(2): 119-134.

25. Kahiluoto, H., M. Kuisma, J. Havukainen, M. Luoranen, P. Karttunen, E. Lehtonen, and M. Horttanainen. 2011. Potential of agrifood wastes in mitigation of climate change and eutrophication - two case regions. Biomass and Bioenergy, 35(5): 1983-1994.

26. Kemp, K., A. Insch, D. K. Holdsworth, and J. G. Knight. 2010. Food miles: Do UK consumers actually care? Food policy, 35(6): 504-513.

27. Kong, D. 2012. Does corporate social responsibility matter in the food industry? Evidence from a nature experiment in China. Food Policy, 37(3): 323-334.

28. Lopez, D. B., M. Bunke, and J. A. B. Shirai. 2008. Marine aquaculture off Sardinia Island (Italy): Ecosystem effects evaluated through a trophic mass-balance model. Ecological Modelling, 21(2): 292-303.

29. Maloni, M. J., and M. E. Brown. 2006. Corporate social responsibility in the supply chain: an application in the food industry. Journal of Business Ethics, 68(1): 35-52.

30. Matos, S., and J. Hall. 2007. Integrating sustainable development in the supply chain: the case of life cycle assessment in oil and gas and agricultural biotechnology. Journal of Operations Management,25 (6): 1083-1102.

31. McKinsey. 2011 The business of sustainability: McKinsey Global Survey results. McKinsey & Company

32. McNeeley, J. A., and S. L. Scherr. 2003. Ecoagriculture: Strategies to Feed the World and Save Biodiversity. London, Covelo Island Press.

33. Mena, C., B. Adenso-Diaz, and O. Yurt. 2011. The causes of food waste in the supplier–retailer interface: Evidences from the UK and Spain. Resources, Conservation and Recycling, 55(6): 648-658.

34. Meul, M., F. Nevens, and D. Reheul. 2009. Validating sustainability indicators: Focus on ecological aspects of Flemish dairy farms. Ecological Indicators. 9(2): 284-295.

35. Negro, O., M., H. and Smits, R. 2007. Explaining the failure of the Dutch innovation system for biomass digestion - A functional analysis. Energy Policy, 35(): 925-938.

36. Nickell, T. D., C. J. Cromey, , Á. Borja, , and K. D. Black. 2009. The benthic impacts of a large cod farm - Are there indicators for environmental sustainability? Aquaculture, 295(3-4): 226-237.

37. Ortiz, M. 2008. Mass balanced and dynamic simulations of trophic models of kelp ecosystems near the Mejillones Peninsula of northern Chile (SE Pacific): Comparative network structure and assessment of harvest strategies. Ecological Modelling, 16(2): 31-46.

38. Pokharel, S., and A. Mutha. 2009. Perspectives in reverse logistics: A review. Resources, Conservation and Recycling, 53(4): 175-182.

39. Pretty, J. N., A. S. Ball, T. Lang, and J. I. L. Morison. 2005. Farm costs and food miles: an assessment of the full cost of the UK weekly food basket. Food Policy. Food policy, 30(6): 1-19.

40. Ridoutt, B. G., P. Juliano, P. Sanguansri, and J. Sellahewa. 2010. The water footprint of food waste: case study of fresh mango in Australia. Journal of Cleaner Production, 18(16-17): 1714-1721.

41. Risku-Norjaa, H., and I. Mäenpääb. 2007. MFA model to assess economic and environmental consequences of food production and consumption. Ecological Economics, 60(4): 700-711.

42. Rodrigues, G. S., I. A. Rodrigues, C. C. A. Buschinelli, and I. Barros. 2010. Integrated farm sustainability assessment for the environmental management of rural activities. Environmental Impact Assessment Review, 30(4): 229-239.

43. Roy, P., Nei, D., Orikasa, T., Xu, Q., Okadome, H., Nakamura, N., and T. Shiina. 2009. A review of life cycle assessment (LCA) on some food products. Journal of Food Engineering, 90(1): 1-10.

44. Sarkis, J., Q. Zhu, and K. Lai. 2011. An organizational theoretic review of green supply chain management literature. International Journal of Production Economics, 130(1): 1-15.

45. Seuring, S. 2012. A review of modeling approaches for sustainable supply chain management. Decision Support Systems, 54(4):1513-1520.

46. Seuring, S., and M. Müller. 2008. From a literature review to a conceptual framework for sustainable supply chain management. Journal of Cleaner Production, 16(15): 1699-1710.

47. Singh, A. 2010. Integrated Reporting: Too Many Stakeholders, Too Much Data? Forbes. Retrieved http://www.forbes.com/sites/csr/2010/06/09/integrated-reporting-too-many-stakeholders-too-much-data/ (accessed August 17, 2012)

48. Teuscher, P., B. Grüninger, and N. Ferdinand. 2006. Risk management in sustainable supply chain management (SSCM): lessons learnt from the case of GMO-free soybeans. Corporate Social Responsibility and Environmental Management, 13(1): 1-10.

49. Vachon, S., and R. D. Klassen. 2006. Extending green practices across the supply chain: the impact of upstream and downstream integration. International Journal of Operations & Production Management, 26(7): 795-821.

50. Van der Vorst, J., S. J. Tromp, and D. J. Van der Zee. 2009. Simulation modelling for food supply chain redesign; integrated decision making on product quality, sustainability and logistics. International Journal of Production Research, 47(23): 6611-6631.

51. Verigna, H. J. 2006. Advanced Techniques for Generation of Energy from Biomass and Waste., ECN publication.

52. Waddock, S., and C. Bodwell. 2004. Managing responsibility: what can be learned from the quality movement? California Management Review, 47(1): 25-37.

53. Walker, H., L. Di Sisto, and D. McBain. 2008. Drivers and barriers to environmental supply chain management practices: Lessons from the public and private sectors. Journal of Purchasing and Supply Management, 14(1): 69-85.

54. Weatherell, A., and J. Allinson. 2003. In search of the concerned consumer UK public perceptions of food, farming and buying local. Journal of Rural Studies, 19(2): 233-244.

55. Welford, R., and S. Frost. 2006. Corporate social responsibility in Asian supply chains. Corporate Social Responsibility and Environmental Management, 13(3): 166-176.

PART II

DAIRY INDUSTRY

CHAPTER 3

An Appraisal of Carbon Footprint of Milk from Commercial Grass-Based Dairy Farms in Ireland According to a Certified Life Cycle Assessment Methodology

DONAL O'BRIEN, PADRAIG BRENNAN, JAMES HUMPHREYS, EIMEAR RUANE, AND LAURENCE SHALLOO

3.1 INTRODUCTION

Grass-based dairy production is a key agricultural industry in some developed nations, particularly Ireland and New Zealand. The dairy sector internationally, however, is also an important source of greenhouse gas (GHG) emissions, responsible for approximately 3% of global emissions (Opio et al. 2013), 10% of Ireland's emissions and up to 20% of New Zealand emissions (Beukes et al. 2010; Deighton et al. 2010). Climate change caused by GHG emissions has become an important political issue. This has led to growing awareness amongst consumers of the potential adverse effects of climate change, which is expected to increase the demand for food products that generate low GHG emissions (Roy et al. 2009). In light of this anticipated demand, major retailers (e.g. Tesco and Wal-Mart) have already put in place sustainability programs to monitor emissions associated with the production of their food products. Therefore, quantifying GHG emissions from all milk-producing nations is becoming a pre-requisite, but especially for countries like Ireland, which export the majority (85%) of their dairy products (CSO 2013).

Simple simulation models of complex biological and technical processes are used to quantify GHG emissions from milk given that direct measurement of agricultural GHG sources (e.g. soils) is difficult and cost prohibitive. The preferred approach to simulate GHG emissions from milk production is the life cycle assessment (LCA). Ideally, LCA should be applied to quantify emissions from all life cycle phases of milk, such as farming, processing, consumption and waste treatment. The methodology, though, is generally applied to the cradle to farm-gate stage of dairy production and, thus, only simulates emissions associated with milk prior to the sale of the product from the farm (Yan et al. 2011). Firstly, this is because of the variation between farming systems, and secondly, most GHG emissions from milk are emitted during the farm stage. For instance, LCA studies of milk to the retail stage report the cradle to farm-gate stage causes 78–95% of GHG emissions from milk (Defra 2007; Gerber et al. 2010).

Usually studies apply a cradle to farm-gate LCA approach to compare the GHG emission intensity or carbon footprint (CF) of milk (kg of GHG/unit of milk) from contrasting dairy farms (O'Brien et al. 2010; Flysjö et al. 2011) or use the approach to assess GHG mitigation strategies (Rotz et al. 2010). The majority of previous research though has been carried out on a limited number (<20–30) of farms (generally not randomly chosen) or based on a single-point modelled national average farm, because significant resources are required to conduct an LCA study of dairy farms (Thomassen et al. 2009). However, researchers are increasingly overcoming this resource challenge, in part through improvements in mobile computer technology (e.g. Thoma et al. 2013). Thus, this implies that there is an increasing scope to complete LCA studies of CF of milk with a large sample of farms (>100).

The major advantage of performing LCA on a large number of dairy farms is that it provides an insight into the variation between CF of milk from commercial farms, which is generally not possible through national average estimates or small-scale studies. In addition, differences between the CF of milk from dairy farms may be related to variation in farm performance (e.g. milk yield/ha), which can facilitate the development of management practices or mitigation strategies to reduce CF of milk (Thomassen et al. 2009). However, apart from DairyCo (2012), previous large-scale European cradle to farm-gate LCA studies of CF of milk have only considered farms where cows do not graze or graze for a short period, usually no longer than 5 months. Thus, strategies suggested to reduce CF of milk from such studies will not be applicable to grass-based farms where cows graze for an extended period (e.g. 9–10 months; O'Brien et al. 2012).

Although several studies have used LCA to quantify CF of milk from grass-based dairy farms, the application of the methodology varies between studies (Yan et al. 2011). To try to overcome this methodology challenge, LCA was applied according to the British Standards Institute (BSI 2011) publicly available specification 2050:2011 (PAS 2050) LCA standard for GHG emissions. In addition, to comply with PAS 2050 certification requirements, an accredited third party (Carbon Trust) assessed all LCA procedures. The goal of this study was to quantify CF of milk from a significant group (>100) of grass-based dairy farms using an LCA method independently accredited to comply with PAS 2050 standards. Additionally, we aimed to determine the factors that caused variation amongst farms' carbon footprints of milk. The study was limited to Ireland, but is also relevant to developed nations where extended grazing of dairy cattle is practiced, given that the application of a certified LCA method to assess emissions from a large sample of grass-based dairy farms is rare.

3.2 MATERIALS AND METHODS

3.2.1 DATA COLLECTION

A random sample of 171 commercial Irish dairy farms was audited between November 2011 and December 2012. The farms sampled were located in the northeast, east, southeast, south and southwest of the country. The farms assessed were therefore a subset of the country's dairy farms and, thus, not representative of the population's CF of milk. The Carbon Trust provided guidance on randomly selecting dairy farms for the region. All data was collected electronically using annual on-farm surveys, electronic feeds of dairy processor milk data and livestock data (DAFM 2011; ICBF 2012).

The Department of Agriculture, Food and the Marine (DAFM) animal identification and movement (AIM) system was used to source bovine data from dairy farms. The AIM system records all births, movements and disposal of farm animals on a daily basis. It is highly reliable and used to meet EU requirements on traceability for bovines (DAFM 2011). Trained auditors carried out farm surveys that corresponded to a 1-year period (fiscal year). Three to 4 hrs of data collection was sufficient to survey a farm, provided source documents, for instance accounting books, were available. Farm auditors collected information on key parameters, for instance, the area of the dairy enterprise, grazing management, concentrate feed, manure management and use of fertiliser, fuel and contractors. In addition, auditors collected farm livestock inventory data, which was cross-checked with national livestock databases (DAFM 2011; ICBF 2012).

Electronic data collected was fed into a farm database to operate an LCA model to calculate the CF of milk from dairy farms. The process described to collect data was externally verified by the Carbon Trust. The Carbon Trust audited farms by cross-checking electronic data collected with farm invoices, milk supply records, auditor measurements and farm livestock records. During this validation procedure, 45 farms were excluded from the analysis because of inadequate or unreliable data (e.g. inconsistency between farmer records of livestock numbers and AIM records). Two more farms were omitted from the evaluation because they had or began to cease producing milk during the period of the study. In total, 124 dairy farms were analysed.

The majority (70%) of farms analysed were grass-based spring calving systems where the aim was to minimize costs through maximizing the proportion of grazed grass in the diet (Kennedy et al. 2005). To achieve this goal, farmers synchronized calving with the onset of grass growth in early or mid-spring and cows remained at pasture from calving until late autumn or early winter. Pasture was usually offered to cows through a rotational grazing system, where cows were offered sections of pasture for 1–2 days or until a specific grazing height was reached (e.g. 4–5 cm) and then moved to a new section. When grass growth exceeded herd feed demand, surplus grass was harvested as grass silage, hay or both and fed to cows indoors from early winter to early spring. Cows were supplemented with purchased concentrate feeds, when grass growth was insufficient to meet herd feed requirements.

Spring calving grass-based dairy systems are the dominant method of producing milk in Ireland (Evans et al. 2004). Thus, the supply pattern of milk is highly seasonal, greatest in May and lowest in January. There is a requirement for some milk to be produced out of season for the fluid milk market and specialty-type products. In order to incentivize farmers to supply milk during the winter period, dairy processors offer farmers a winter contract with a milk price bonus to cover the extra costs of production (Fitzgerald et al. 2004). To meet this requirement, a minority of farms analysed calved cows in autumn or throughout the year. Thus, these farms offered milking cows concentrate feed over the winter and, in addition to grass silage, fed maize silage, whole crop silage or both. Farms that milked cows throughout the winter, however, also aim to graze cows for an extended period (6–8 months) to minimize costs, particularly in late autumn and early spring (Fitzgerald et al. 2004).

3.2.2 GREENHOUSE GAS SIMULATION

A cradle to farm-gate attributional LCA model developed by O'Brien et al. (2010) was used to simulate annual GHG emissions from dairy farms. Thus, all sources of GHG emissions associated with dairy production until milk leaves the farm were simulated, including off-farm GHG sources such as fertiliser production (Fig. 1). The model calculated GHG emissions by combining information from the farm database, mentioned previously, with literature emission algorithms. However, the LCA model procedures and literature emission factors were adapted based on farm data availability and recommendations of the Carbon Trust. In addition, fluorinated gases (F-gases) from refrigerant loss were added to the model as a source of GHG emissions and estimated based on farm service records of cooling equipment.

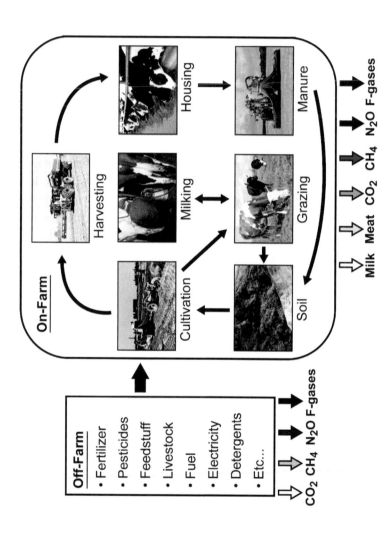

FIGURE 1: An illustration of the major sources of on- and off-farm greenhouse gas emissions, carbon dioxide (CO_2), methane (CH_4), nitrous oxide (N_2O) and fluorinated gases (F-gases), quantified using a cradle to farm-gate life cycle assessment model

The main on-farm sources of GHG emissions quantified by the LCA model are summarized in Table 1. Enteric methane (CH_4) emissions were estimated by firstly computing animal feed intakes. To ensure animal feed intakes were realistic, they were calculated to fulfil net energy requirements for milk production, maintenance, pregnancy and body weight change (Jarrige 1989). Where possible, information directly acquired on animal and feed variables were used to validate feed intakes. However, this was not possible for all variables, for instance energy values of forages. Thus, data from literature sources were also used (O'Mara 1996, 2006). Enteric CH_4 emissions were estimated as a function of intake, using a fixed factor of 6.5% of gross energy intake (GEI) when cattle grazed grass (Duffy et al. 2011b). However, when the diet comprised of only silage and concentrate, a regression equation from Yan et al. (2000) was used.

TABLE 1: Key emission factors applied in a cradle to farm-gate life cycle assessment model to quantify on-farm greenhouse gas (GHG) emissions, ammonia emissions and nitrate loss

Emission and source	Emission factor	Unit	References
Carbon dioxide (CO_2)			
Lime	$0.12 \times$ lime application	kg/kg lime	IPCC (2006)
Urea	$0.20 \times$ urea application	kg/kg urea	IPCC (2006)
Diesel	$2.63 \times$ diesel use	kg/l	IPCC (2006)
Gasoline	$2.30 \times$ gasoline use	kg/l	IPCC (2006)
Kerosene	$2.52 \times$ kerosene use	kg/l	IPCC (2006)
Liquefied petroleum gas (LPG)	$1.49 \times$ LPG use	kg/l	IPCC (2006)
Methane (CH_4)			
Enteric fermentation			
Dairy cow and heifer (housing)	DEI $\times (0.096 + 0.035 \times S$ DMI$/$T DMI$) - (2.298 \times FL^{-1})$	MJ/day	Yan et al. (2000)
Dairy cow and heifer (grazing)	$0.065 \times$ GEI	MJ/day	Duffy et al. (2011b)
Manure storage and excretion on pasture	Manure VS excreted $\times 0.24 \times 0.67 \times MS^a \times MCF^b$	kg/year	IPCC (2006); Met (2013)
Nitrous oxide (N_2O)			
Slurry storage	$0.005 \times$ slurry N stored	kg/kg N	IPCC (2006)
Solid manure storage	$0.005 \times$ solid manure N stored	kg/kg N	IPCC (2006)

TABLE 1: *Cont.*

Emission and source	Emission factor	Unit	References
Manure excreted on pasture	$0.02 \times N$ excreted on pasture	kg/kg N	IPCC (2006)
Synthetic N fertiliser	$0.01 \times N$ fertiliser	kg/kg N	IPCC (2006)
Manure application	$0.01 \times$ (manure N applied $-$ N storage loss)	kg/kg N	IPCC (2006)
Crop residues	$0.01 \times N$ crop residues	kg/kg N	IPCC (2006)
Nitrate leaching	$0.0075 \times N$ leached	kg/kg NO_3 $-$-N	IPCC (2006)
Ammonia (NH_3) re-deposition	$0.01 \times$ sum of NH_3 loss	kg/kg NH_3-N	IPCC (2006)
Ammonia (NH_3-N)			
Housing	11–38[c]	g/luuk per day	Duffy et al. (2011a)
Slurry storage	$(2-4)^d \times$ area slurry store	g/m² per day	Duffy et al. (2011a)
Solid manure storage	$94 \times$ area solid manure store	g/m² per year	Duffy et al. (2011a)
Slurry application	$(0.15-0.59)^e \times$ TAN in slurry spread	kg/kg TAN	Duffy et al. (2011a)
Solid manure application	$0.81 \times$ TAN in solid manure spread	kg/kg TAN	Duffy et al. (2011a)
Grazing cattle	$0.2 \times N$ excreted on pasture	g/luuk per day	Duffy et al. (2011a)
Synthetic N fertiliser	$(0.08-0.23)f \times N$ fertiliser applied	kg/kg N	Duffy et al. (2011a)
Nitrate (NO_3 $-$-N)			
N leaching	$0.1 \times$ (N applied $-NH_3$ loss $-N_2O$ loss)	kg/kg N	Duffy et al. (2011b)

DEI digestible energy intake, S DMI silage dry matter intake, T DMI total dry matter intake, FL feeding levels above maintenance energy requirement, GEI gross energy intake, VS volatile solids, luuk UK livestock unit (equivalent to 500 kg body weight), TAN total ammoniacal nitrogen [a]MS = percentage of manure volatile solids managed in a specific storage system or percentage of manure excreted on pasture. [b]MCF = methane conversion factor for manure volatile solids managed in a particular storage system or excreted on pasture. The MCF values were 0.01 for manure excreted on pasture, 0.02 for solid manure system (dry matter (DM) >20%), 0.66 for lagoon system and 0.17 for slurry system without a surface crust or 0.11 with a surface crust [c]Dependent on the age of the animal [d]Dependent on manure storage facility [e]Dependent on DM of slurry and season of application [f]Dependent on fertiliser compound

Methane emissions from manure were estimated according to the Intergovernmental Panel on Climate Change IPCC (2006) guidelines as a proportion of the maximum CH_4 potential (B_o) of manure volatile solids (VS). Based on O'Mara (2006), manure VS excretion was calculated by multiplying animal organic matter (OM) intake by the indigestible OM component of the diet. The quantity of manure VS requiring storage was computed based on the number of days animals spent housed. The B_o of manure VS for dairy cattle was based on Duffy et al. (2011b). The proportion of the B_o that was emitted from manure VS was computed using specific CH_4 conversion factors for different manure storage systems, but for manure deposited on pasture, a default factor (1%) was used (IPCC 2006). The CH_4 conversion factors for manure storage systems were selected assuming an annual average ambient temperature of 10°C for Ireland (Met 2013).

Emissions of nitrous oxide (N_2O) from manure were derived after calculating N excretion, which was estimated as the difference between total N intake and N output in meat and milk. Direct N_2O emissions from manure storage were estimated using manure storage-specific emission factors. After subtraction of N losses during housing and storage, N_2O emissions from manure spreading were estimated as 1% of N applied (IPCC 2006). In addition, this factor was used to estimate N_2O emissions from synthetic fertiliser spreading, crop residues and from soil mineralisation following land use change. Nitrogen inputs from crop residues and soil mineralisation were computed based on the IPCC (2006) guidelines. A greater N_2O emission factor was estimated for manure excreted by grazing cattle (2% of N excreted), compared to manure or fertiliser spreading, given that urine deposited by grazing cattle can cause large N losses (Van Groenigen et al. 2005).

Indirect N_2O emissions from re-deposition of volatilized NH_3 were computed as 1% of NH_3 emitted from fertiliser and manure (IPCC 2006). Volatilization of NH_3 from synthetic fertiliser application was estimated using N loss factors from Hyde et al. (2003). A mass flow approach and emission factors from Hyde et al. (2003) were used to estimate NH_3 emitted during cattle housing, manure storage and spreading. Ammonia loss from manure excreted by grazing cattle was estimated as 20% of N excreted on pasture (IPCC 2006). Indirect N_2O emitted from leaching of N was estimated as 0.75% of N leached (IPCC 2006). A default leaching factor from Duffy et al. (2011b) was used to estimate nitrate leaching. Nitrogen available for leaching was quantified by subtracting N removed in products and N lost directly to the atmosphere from total N inputs (manure, soil mineralisation, fertiliser and crop residues).

On-farm emissions of carbon dioxide (CO_2) from fossil fuels, lime and urea were estimated using the IPCC (2006) guidelines. Short-term biogenic sources and sinks of CO_2 such as animals, crops and manure were considered to be neutral with respect to GHG emissions given that the IPCC (2006) assume all C absorbed by animals, crops and manure to be quickly released back to the atmosphere through respiration, burning and decomposition. Agricultural soils also have the potential to emit or sequester CO_2 (Rotz et al. 2010). However, to comply with PAS 2050, C sequestration was not included for permanent pasture. This was because the standard follows the IPCC (2006) recommendation that soil's ability to store or lose C reaches equilibrium after a fixed period (20 years). Thus, land use change emissions were also restricted to this period and estimated as 6.7–7.0 t CO_2/ha per annum when permanent on-farm grassland was converted to cropland (Carbon Trust 2013).

Data acquired on external farm inputs (e.g. diesel and pesticides) were used primarily with emission factors (Table 2) from the Carbon Trust (2013) to estimate off-farm GHG emissions. For electricity generation and some other sources though, it was more appropriate to use emission algorithms from national literature sources. When emissions from an external farm input could not be estimated via the Carbon Trust or national literature, Ecoinvent (2010) was used. The production of specific imported feeds, for instance Malaysian palm kernel, was estimated to cause land use change emissions by computing the average land use change emissions for that crop in that country (Carbon Trust 2013). Directly attributing land use change emissions to a crop conforms to the method used by the Food and Agriculture Organization (FAO) to estimate GHG emissions from milk (Opio et al. 2013).

TABLE 2: Key emission factors used in a cradle to farm-gate life cycle assessment model for quantification of off-farm greenhouse gas emissions in kilograms of CO_2 equivalent

Item	Emission factor	Reference
Electricity, kWh	0.60	Howley et al. (2011)
Diesel, l	0.41	Ecoinvent (2010)
Gasoline, l	0.57	Ecoinvent (2010)
Kerosene, l	0.39	Ecoinvent (2010)
Liquefied petroleum gas, l	0.30	Ecoinvent (2010)
Ammonium nitrate, kg N	7.11	Carbon Trust (2013)
Urea, kg N	3.07	Ecoinvent (2010)
Lime, kg	0.15	Carbon Trust (2013)
P fertiliser, kg P_2O_5	1.86	Carbon Trust (2013)
K fertiliser, kg K_2O	1.77	Carbon Trust (2013)
Refrigerant, kg	11.00–393.00	Little (2002)
Detergent, kg active ingredient	0.11–1.03	Ecoinvent (2010)
Pesticide, kg active ingredient	7.37	Carbon Trust (2013)
Barley, kg dry matter (DM)	0.35	Carbon Trust (2013)
Corn grain, kg DM	0.45	Carbon Trust (2013)
Citrus pulp, kg DM	0.06	Carbon Trust (2013)
Corn gluten, kg DM	0.34	Carbon Trust (2013)
Molasses, kg DM	0.15	Carbon Trust (2013)
Rapeseed meal, kg DM	0.40	Carbon Trust (2013)
South America soybean meal, kg DM	11.65	Carbon Trust (2013)
USA soybean meal, kg DM	0.32	Carbon Trust (2013)
South America soybean hulls, kg DM	0.28	Ecoinvent (2010)
		Carbon Trust (2013)
USA soybean hulls, kg DM	0.01	Ecoinvent (2010)
		Carbon Trust (2013)
Compound concentrate, 16% crude protein[a] (CP), kg DM	0.34	Ecoinvent (2010)
		Carbon Trust (2013)
Compound concentrate, 20% CP[b], kg DM	1.98	Ecoinvent (2010)
		Carbon Trust (2013)
Compound concentrate, 32% CP[c], kg DM	3.37	Ecoinvent (2010)
		Carbon Trust (2013)

aConcentrate formulation on a DM basis: USA soy hulls 18%, USA dried distillers grains 17%, USA citrus pulp 17%, USA corn gluten feed 8%, French rapeseed meal 8%, German corn grain 6%, Cuban molasses 5%, Irish barley 5%, Malaysian palm kernel meal 4%, Irish wheat feed 4%, vegetable oil 3%, Irish lime 3%, minerals and vitamins 2% bConcentrate formulation on a DM basis: Irish wheat feed 17%, USA soy hulls 16%, French rapeseed meal 15%, Brazilian soybean meal 14%, German corn grain 7%, Cuban molasses 6%, Malaysian palm kernel meal 6%, USA dried distillers grains 4%, USA citrus pulp 4%, French sunflower meal 4%, Irish lime 3%, vegetable oil 2%, minerals and vitamins 2% cConcentrate formulation on a DM basis: Brazilian soybean meal 26%, French rapeseed meal 24%, USA dried distillers grains 12%, Malaysian palm kernel meal 8%, USA soy hulls 7%, Cuban molasses 6%, USA corn gluten feed 5%, French sunflower meal 4%, Irish lime 3%, vegetable oil 3%, minerals and vitamins 2%

3.2.3 QUANTIFICATION AND ACCREDITATION OF CF OF MILK FROM DAIRY FARMS

On- and off-farm GHG emissions were converted to CO_2 equivalent (CO2-eq) emissions using the IPCC (2007) guidelines' global warming potential (GWP) factors, which have been revised (IPCC 2013) and summed to compute dairy farms' annual CO_2-eq emissions. The GWP factors for key GHG emissions were 1 for CO_2, 25 for CH_4 and 298 for N_2O, assuming a 100-year time horizon. The CF of milk from dairy farming was estimated by firstly allocating GHG emissions between farm outputs, milk, crops, manure and meat from culled cows and surplus calves. Where possible, PAS 2050 recommendations to avoid allocation of GHG emissions between products were applied. However, this was only achieved for exported crops by constraining the LCA model to quantify emissions from crops grown for dairy cattle.

When allocation was required, GHG emissions were allocated between dairy farm products based on their economic value. The economic method of allocation was used instead of alternative methods, e.g. the physical allocation approach recommended by the IDF (2010), because economic allocation is preferred by PAS 2050 when allocation cannot be avoided. The economic value of milk and meat was estimated using 5-year average prices from 2007 to 2011 (CSO 2013). Cattle manure exported from dairy farms had no economic value. Thus, GHG emissions were allocated to cattle manure based on the site of storage and application. Emissions from the transport of exported manure were attributed to the importer. Greenhouse gas emissions attributed to milk were expressed per kilogram of fat and protein-corrected milk (FPCM) to quantify the CF of milk from dairy farms. An algorithm from the International Dairy Federation (IDF 2010) was used to estimate FPCM as 4% fat and 3.3% true protein.

The LCA model estimate of CF of milk was tested by the Carbon Trust to certify compliance with PAS 2050. Research data from Teagasc (2011) on poor, average and high performing grass-based Irish dairy farms were initially used to evaluate the model. Subsequently, research data from Olmos et al. (2009) was used to test the model's ability to estimate CF of milk from a high input farm where cows did not graze. The LCA model was also tested with data from commercial farms. This was achieved by randomly selecting 12 grass-based farms from the study sample.

To achieve accreditation, non-conformities identified by the Carbon Trust between the LCA model and PAS 2050 were addressed. Non-conformities that had a non-material impact on the CF of milk were either justified through research data or changed based on advice of the Carbon Trust. PAS 2050 certification was granted when all non-conformities were addressed or where the effect of all unresolved non-material non-conformities on the CF of milk was $<\pm5\%$. To determine if the LCA model CF result was within the $<\pm5\%$ threshold, all changes recommended by the Carbon Trust were applied to any model non-conformity with PAS 2050.

3.2.4 STATISTICAL ANALYSES

The Statistical Analysis Systems (SAS) Institute software package (SAS 2008) was used to evaluate relationships between farm characteristics (e.g. herd size) and CF of milk. The strength of the relationship between CF of milk and individual farm characteristics was measured using the Pearson correlation where the data was normally distributed and the Spearman rho correlation for non-normal data. Correlations were also performed amongst farm characteristics, where a farm variable was correlated to CF of milk. Least squares regression analysis was used to evaluate the associations between individual farm variables and CF of milk. The equality of the variances of residuals of regression models was checked for normality visually and the Shapiro-Wilk test was used (P<0.05). The significance of regression coefficients was assessed using the t statistic.

Non-linear terms were added to regression models based on a visual assessment of the distribution of the residuals. The stepwise multiple regression procedure of SAS (2008) was used to determine whether the addition of non-linear terms made a significant contribution (P<0.05) to a regression equation. The procedure was also used to develop predictive models by assessing the relationship between the CF of milk and all farm characteristics listed in Tables 3 and 4. Variance inflation factors and condition indices were applied in SAS (2008) to check for multicollinearity within multiple regression models. Regression coefficients with variance inflation factors >10 were omitted.

TABLE 3: Weighted means, standard deviations (SD), coefficients of variations (CV), minimum, maximum and lower and upper 10 percentiles of key farm characteristics of 124 Irish dairy farms

Farm characteristic	Mean	SD	CV (%)	Min	Lower 10%	Upper 10%	Max
Farm size, ha	51	27	52	18	30	75	245
Stocking rate, LU/ha	2.10	0.46	22	1.30	1.50	2.60	2.93
Cows, n	91	55	60	29	44	149	468
Annual FPCM yield/cow	5,380	834	16	3,163	4,433	6,483	8,071
Annual FPCM yield/ha	9,494	2,534	27	4,529	6,435	12,832	15,085
Fat,%	3.89	0.17	4	3.54	3.71	4.13	4.31
Protein,%	3.38	0.10	3	3.10	3.26	3.51	3.73
Lactose,%	4.70	0.05	1	4.48	4.64	4.76	4.84
Replacement rate,%	19	8	43	7	9	28	36
Grazing season, days	245	23	9	184	214	274	280
Concentrate, kg DM/cow	643	310	48	224	310	1,062	1,552
Purchased forage, kg DM/cow	118	202	171	0	0	337	1,197
Total feed, kg DM/cow	5,630	549	10	4,270	4,961	6,350	6,444
On-farm N fertiliser, kg/ha	172	54	32	95	112	255	307
Farm-gate N surplusa, kg/ha	150	53	36	70	87	231	240

TABLE 3: *Cont.*

Farm characteristic	Mean	SD	CV (%)	Min	Lower 10%	Upper 10%	Max
Farm N efficiencyb,%	27	7	25	14	20	37	39
Electricity, kWh/cow	253	98	38	57	152	388	409
On-farm fuel, l/ha	97	15	15	41	78	116	121

LU livestock unit (equivalent to the average annual N excretion of an Irish dairy cow); FPCM fat and protein-corrected milk, where milk was standardised to 4% fat and 3.3% true protein per kilogram; DM dry matter [a]N imports–N exports passing in or out through the farm-gate [b]Farm N exports/farm N imports

TABLE 4: Correlations (r) between various farm characteristics and carbon footprint (CF) of milk

Farm characteristic	CF of milk	P value
Farm size, ha	0.07	NS
Cows, n	−0.04	NS
Cows/ha	−0.33	***
Stocking rate, LU/ha	−0.32	***
Total farm milk production, t FPCM	−0.17	NS
Annual FPCM yield/cow	−0.48	***
Annual FPCM yield/ha	−0.55	***
Fat,%	−0.24	**
Protein,%	−0.31	***
Lactose,%	−0.18	NS
Replacement rate,%	0.03	NS
Grazing season, days	−0.45	***
Concentrate, kg DM/cow	0.15	NS
Purchased forage, kg DM/cow	0.20	*
Total feed, kg DM/cow	−0.24	**
On-farm forage use, kg/ha	−0.48	***
On-farm N fertiliser, kg/ha	0.09	NS
Farm-gate N surplus[a], kg/ha	0.28	**
Farm N efficiency[b],%	−0.66	***
Electricity, kWh/cow	0.10	NS
On-farm fuel, l/ha	0.06	NS

Fat and protein-corrected milk standardised to 4% fat and 3.3% true protein per kilogram LU livestock unit (equivalent to the annual N excretion of an average Irish dairy cow), DM dry matter, NS not significant *P < 0.05; **P < 0.01; ***P < 0.001 ^aN imports–N exports passing in or out through the farm-gate ^bFarm N exports/farm N imports

3.3 RESULTS

3.3.1 GENERAL FARM CHARACTERISTICS

The weighted mean, standard deviation (SD), coefficient of variation (CV) and range for various farm characteristics of 124 dairy farms are shown in Table 3. On average, dairy farms were 51 ha in size, stocked at 2.1 livestock units (LU)/ha and produced 489 t of FPCM. The mean herd size was 91 cows and the replacement rate averaged 19% (SD=8%) across farms. Cows spent the majority of the year at pasture (mean of 245 grazing days, SD=23). Thus, their diet was mainly composed of grazed grass (mean 63%; SD=7%). Concentrate input averaged 643 kg DM/cow (SD=310) across farms, and the mean quantity of forage purchased was 118 kg DM/cow (SD=202). On-farm N fertiliser application averaged 172 kg N/ha (SD=54) and the mean farm-gate N surplus (N imports–N exports) was 150 kg N/ha (SD=53). On-farm electricity consumption averaged 47 kWh of electricity/t of FPCM (SD=18) and fuel use was 97 l of fuel/ha (SD=15).

3.3.2 CERTIFIED CF OF MILK FROM IRISH DAIRY FARMS

All non-conformities identified between the LCA model and PAS 2050 during evaluation of the CF of milk from dairy farms were addressed. Generally, non-conformities were resolved by applying changes recommended by the Carbon Trust, but for some non-conformities considered non-material (e.g. simulation of manure excretion by cattle), no revisions were undertaken. Overall, the cumulative effect of non-material non-conformities was <1%. Thus, CF of milk estimated by the LCA model was accredited to comply with PAS 2050.

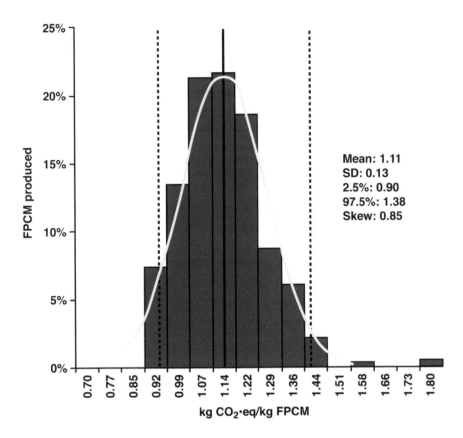

FIGURE 2: Histogram and expected normal distribution (bell-shaped line) of carbon footprint of fat and protein-corrected milk (FPCM), weighted by farm FPCM production for 124 dairy farms in CO_2 equivalent (CO_2-eq). The middle vertical solid line indicates the mean value and the dotted vertical lines to the left and right of the mean indicate the interval between the lower 2.5 % and upper 97.5 % of values. The mean carbon footprint of milk and standard deviation (SD) are also shown

On average, 90% of dairy farms' GHG emissions were allocated to milk production. The mean certified CF of milk to the farm-gate, weighted by farm milk production, was 1.11 kg of CO_2-eq/kg of FPCM. The SD was within ±0.13 kg of CO_2-eq of the weighted mean. Figure 2 shows across the 124 dairy farms that there was significant variability between farms' CF of milk, ranging from 0.87 to 1.72 kg of CO_2-eq/kg of FPCM. On average, CF of milk mainly consisted of CH_4 (47%) and N_2O (34%) emissions, followed by emissions of CO_2 (19%) and F-gases (<0.5%). Approximately, 80% of the CF of milk to the farm-gate was caused by GHG emissions generated directly on-farm.

The largest individual on-farm contributor to CF of milk was CH_4 from enteric fermentation (44%), which averaged 113 kg CH_4/cow. Nitrous oxide from manure excreted by grazing cattle was the next largest on-farm contributor (17%) to CF of milk, followed by N_2O and CO_2 emissions from fertiliser application (8%), CH_4 and N_2O emissions from manure storage and spreading (6%) and CO_2 emissions from fuel consumption and lime (5%). The remainder of the CF of milk comprised of off-farm GHG emissions from fertiliser manufacture (10%), concentrate production (7%) and energy generation (3%). The majority of GHG emissions from concentrate production were caused by CO_2 emissions from land use change (57%).

3.3.3 ASSOCIATIONS BETWEEN FARM CHARACTERISTICS AND CF OF MILK

Significant (P<0.05) associations between farm characteristics and CF of milk were linear. Farm N efficiency had the strongest negative correlation with CF of milk (r=−0.66; Table 4), which regression analysis found decreased CF of milk by 13 g of CO_2-eq (standard error (se)=1) for each 1% increase in N efficiency. Apart from total farm milk production, measures of milk yield were moderately negatively correlated to CF of milk, but the correlation was stronger when evaluated per hectare (r=−0.55) rather than per cow (r=−0.48). Milk yield/cow had a similar negative correlation with CF of milk as homegrown forage utilisation per hectare, but a slightly stronger correlation than the length of the grazing season (r=−0.45). Moderate to weak negative correlations (r=−0.24 to −0.32) occurred between

CF of milk and milk fat content, milk protein content, total feed intake/cow and stocking rate.

Farm N surplus per hectare and purchased forage/cow were the only farm attributes positively correlated to CF of milk, but the correlations were weak ($r = 0.20$ to 0.28). Annual replacement rate, farm size and herd size were not associated with CF of milk. However, stepwise multiple regression (Table 5) showed that CF of milk could be reasonably well explained ($R^2 = 0.75$) by farm N efficiency, the length of the grazing season, milk yield/cow and annual replacement rate. Of these farm attributes, the procedure found that the majority of variation in CF of milk amongst farms was explained by farm N efficiency.

TABLE 5: Regression coefficients (b), associated standard errors (SE), significance of regression coefficient and coefficient of determination (R^2) estimated in a stepwise multiple regression model of carbon footprint of milk

Parameter	b	SE	P value	R^2
Intercept	2.19	0.08	***	–
N efficiency[a],%	−0.98	0.10	***	0.46
Grazing season, days	-2.16×10^{-3}	2.71×10^{-4}	**	0.61
FPCM yield/cow	-7.76×10^{-5}	9.87×10^{-6}	**	0.70
Replacement rate,%	3.37×10^{-3}	9.43×10^{-4}	*	0.75

Fat and protein-corrected milk standardised to 4% fat and 3.3% true protein per kilogram
$*P < 0.05$; $**P < 0.01$; $***P < 0.001$ [a]Farm N exports/farm N imports

3.4 DISCUSSION

The application of LCA according to PAS 2050 highlighted a significant variation in the CF of milk amongst grass-based dairy farms. The differences between CF of milk were mainly related to variation in aspects of farm performance and characteristics. Thus, this indicates that grass-based dairy producers can mitigate the CF of milk by adopting management practices that improve efficiency and performance. Furthermore, the study suggests that the goals of maintaining farm profitability and reducing CF of milk are not contradictory, given that previous studies of grass-based research farms report that improving farm productivity increases profit (Lovett et al. 2008; O'Brien et al. 2010). However, no single farm attribute accounted for the majority of variation between farms' CF of milk. Thus, the study implies, similar to Beukes et al. (2010), that a suite of farm practices are required to increase efficiency and thus reduce carbon footprint of milk.

3.4.1 COMPARISONS WITH NATIONAL AND INTERNATIONAL STUDIES

Relative to previous single-point estimates of the national average CF of Irish milk (Casey and Holden 2005; Lovett et al. 2008; Teagasc 2011), our mean result for commercial dairy farms was lower by 7–16%. This was primarily because we considered a subset of dairy farms that were on average moderately more productive (e.g. higher milk yield/ha) than the national average farm. For example, compared to the performance of the average Irish dairy farm for 2010 and 2011 (Teagasc 2011), the farms analysed produced more FPCM/cow (5,380 versus 5,058 kg) and per hectare (9,494 versus 8,982 kg) and fed approximately 276 kg DM less concentrate/cow. In addition, the mean period cows spent at pasture was longer than the national average, which Lovett et al. (2008) reported and the present results indicate to mitigate GHG emissions from grass-based dairy production. However, relative to a recent European estimate of the average CF of Irish milk by Leip et al. (2010), our results were approximately 10% higher, despite the greater efficiency of the dairy farms we assessed. The anomaly, though, was simply due to different LCA modelling assumptions.

For instance, Leip et al. (2010) included carbon sequestration by permanent grassland, which PAS 2050 excludes based on the IPCC (2006) guidelines. Recent studies report that permanent grasslands are an important long-term carbon sink (Soussana et al. 2007, 2010). Adopting the same carbon sequestration assumptions in the present study as Leip et al. (2010) reduced the mean CF of milk to 0.95 kg of CO_2-eq/kg FPCM, which is 5% lower than the estimate of Leip et al. (2010) of the average

CF of Irish milk. Thus, as more carbon sequestration data becomes available, PAS 2050 may need to be revised to include this sink. The large influence that alternative modelling choices (e.g. allocation methodologies) have on the CF of milk and meat is a well-documented LCA issue (Flysjö et al. 2011). Thus, until a harmonised LCA approach such as PAS 2050 is widely adopted internationally, direct comparisons between CF studies of milk are of limited value. Consequently, the FAO is leading a livestock environmental assessment and performance partnership (LEAP 2014) to develop global LCA guidelines for a wide array of environmental impacts from livestock. The draft LEAP guidelines generally adopt the same principles as PAS 2050, but are specific for animal supply chains.

Although methodological variations often partially explain differences between LCA studies, cautious comparisons can provide an indication of the validity of LCA results and are useful in understanding the potential to mitigate CF of milk. Our results were similar to pan-European and global LCA studies, which showed that grazing systems of developed nations have the lowest CF of milk (Leip et al. 2010; Hagemann et al. 2012; Opio et al. 2013). However, such studies generally only assess the national or regional average situation and do not assess the variation in CF of milk within a region. Several reports suggest that this variation is significant. For instance, DairyCo (2012) showed that even though the average CF of UK milk was in the lower range of literature estimates, substantial differences were found amongst 415 grazing and non-grazing dairy systems' CF of milk (0.83–2.81 kg of CO_2-eq/l of fat corrected milk). A similar variability has also been reported for CF of Dutch milk (Thomassen et al. 2009) and USA milk (Thoma et al. 2013) where commercial non-grazing systems were predominately assessed and by an Australian evaluation of 140 grazing and non-grazing dairy systems (Dairy Australia 2012). In addition, the results of these studies show that farmers of developed nations can reduce CF of milk via management changes, e.g. DairyCo (2012) reported that reducing concentrate supplementation of cows decreased CF of milk of dairy farms. Thus, this indicates that global or European single-point estimates of national CF of milk may underestimate the scope to mitigate GHG emissions.

Generally, our analysis agreed with large-scale LCA studies (e.g. Thomassen et al. 2009; Dairy Australia 2012) regarding the range of farms' CF of milk and the main sources of dairy farms' GHG emissions (e.g. enteric CH4). However, unlike LCA studies of similar scale, the farms we considered were not nationally representative and, as discussed marginally, more productive than the national average dairy farm. Consequently, the variability amongst farms' CF of milk was lower than previous large-scale LCA studies (e.g. DairyCo 2012). There was, however, a significant variation amongst key sources of dairy farms' GHG emissions, particularly N2O emissions from manure and emissions associated with on-farm fertiliser use. The variability of these GHG sources was similar to previous reports by Dairy Australia (2012) and Thoma et al. (2013). Thus, these results suggest that there is potential to mitigate GHG emissions of grass-based dairy systems, which can be realised in part by adopting new technologies, but also via changes in farm management practices.

3.4.2 IMPACT OF FARM EFFICIENCY AND PERFORMANCE ON CF OF MILK

Similar to Casey and Holden (2005), Christie et al. (2011) and Yan et al. (2013b), numerous measures of farm efficiency and performance were correlated to CF of milk from grass-based farms. Congruous with Yan et al. (2013a), the farm attribute that had the strongest association with CF of milk was farm N efficiency (farm N imports/farm N exports). Key management practices that improve N efficiency of grass-based farms included adoption of white clover, greater utilisation of manure and better timing of manure and fertiliser application to grass growth. These practices reduced surplus N or the requirement for N fertiliser, which increased farm N efficiency. However, reducing N use on an area or animal basis rather than per unit of milk had little or no influence on CF of milk, which agrees with similar analysis by Olesen et al. (2006). Therefore, this may explain why several studies have generally not found or reported weak relationships between farm N use and CF of milk, given that most studies only assess farm N use per hectare (Dairy Australia 2012; DairyCo 2012).

Grass-based farms that were not efficient from an N perspective, however, did not necessarily have a high CF of milk. The main explanation for this was improving farm N efficiency primarily reduced emissions associated with N fertiliser, but had no effect on enteric CH_4 emissions, which were the main components of farms' CF of milk. Therefore, farms that emitted lower enteric CH_4 emissions/unit of milk than more N efficient farms could achieve a lower CF of milk. Increasing milk yield/cow and reducing the annual herd replacement rate were the primary farm attributes that mitigated enteric CH_4 emissions. Improving these attributes increased feed conversion efficiency (total feed DM/kg of FPCM), which in agreement with Rotz et al. (2010) was the main determinant of enteric CH_4 emissions/unit of milk. Increasing cow genetic merit via artificial insemination was the main management practice available to farmers to increase cow milk performance. Furthermore, the practice could also be used to improve cow fertility or health, which reduces the requirement for replacement heifers. However, this was achievable only using sires of a total genetic merit index, e.g. Irish economic breeding index (EBI), which are bred to increase cow performance, fertility and health.

Improving milk yield/cow also reduced CF of milk, but only caused minor reductions in GHG emissions from N inputs, e.g. fertiliser. In addition, similar to Ramsbottom et al. (2012), higher milk yield/cow was associated with greater concentrate feeding, which increased GHG emissions. Consequently, milk yield/cow was not as influential in determining CF of milk as farm N efficiency. Improving milk yield per hectare rather than per cow had less of an influence on feed conversion efficiency, but tended to cause a greater improvement in N efficiency. Furthermore, the measure increased homegrown forage utilisation per hectare, which in agreement with previous studies, reduced emissions from concentrate (Thomassen et al. 2009; O'Brien et al. 2010). Thus, improving milk yield per hectare had a slightly greater mitigating influence than increasing milk output/cow on CF of milk. Stocking rate was a key determinant of milk yield per hectare. However, stocking rate had little or no effect on feed conversion efficiency. Thus, this explained the lower influence milk yield per hectare had on feed conversion efficiency relative to milk yield/cow and the weak negative association between stocking rate and CF of milk.

Homegrown forage utilisation per hectare was primarily determined by stocking rate, but had a stronger association with CF of milk. This was because the farm attribute, unlike stocking rate, tended to be positively associated with the grazing season. Extending the grazing season shortened the winter housing period, which similar to Schils et al. (2005, 2007) reduced GHG emissions from on-farm fuel use, manure storage and spreading. In addition, the management practice increased grazed grass utilisation per hectare, which decreased GHG emissions from concentrate and, in contrast to the results of Schils et al. (2007), reduced enteric CH_4 emissions. However, the main forage fed to cows indoors in the study of Schils et al. (2007) was maize silage, whereas in the present study, grass silage was mainly fed, which yields more enteric CH_4 than grazed grass (Robertson and Waghorn 2002). Therefore, our results showed, unlike Schils et al. (2005, 2007), that extending the grazing season reduced CF of milk, but the negative association was only moderate. This was because the farm practice also increased N_2O emissions from manure excreted on pasture by grazing cattle.

No single farm attribute analysed explained the majority of variation between the CF of milk from grass-based farms. As a result, similar to previous studies, improving farm measures in isolation had a minor mitigating effect on CF of milk (Lovett et al. 2008; Dairy Australia 2012; DairyCo 2012). Nevertheless, CF of milk from grass-based farms could be well predicted using just farm N efficiency, the length of the grazing season, milk yield/cow and the annual herd replacement rate. Thus, it may not be necessary to conduct detailed on-farm surveys to estimate CF of milk. However, to validate the prediction capabilities of these farm attributes, a new independent audit would need to be conducted.

As discussed, the farm attributes that predicted CF of milk were influenced by farm management practices, e.g. stocking rate, which varied to a similar or greater extent as GHG emissions between farms. Therefore, in order to reduce carbon footprint of milk, grass-based dairy farmers need to implement a suite of management practices to simultaneously improve farm N efficiency, the length of the grazing season, milk yield/cow and the annual herd replacement rate. Farm management practices that positively influenced these key determinants of CF of milk and did not negatively interact were improving cow total genetic merit via artificial insemination, extending the length of the grazing season, reducing concentrate feeding, increasing stocking rate and reducing fertiliser N use per hectare by increasing the proportion of manure applied in spring. Furthermore, implementing these practices improves farm productivity. This increases the economic viability of grass-based farms (Lovett et al. 2008; Beukes et al. 2010; O'Brien et al. 2014) which is a key consideration when changing management practices. Therefore, this implies that grass-based dairy farmers can potentially significantly mitigate CF of milk and maintain farm profitability.

3.5 CONCLUSIONS

Independent certification of an LCA model according to PAS 2050 for a large group of grass-based dairy farms provides a verifiable approach to quantify CF of milk at a farm or national level. The application of the certified model in this study showed that the mean CF of milk from a subset of grass-based Irish farms was in the lower range of literature estimates. However, comparisons with previous LCA studies that were not certified to the PAS 2050 standard were affected by inconsistent modelling assumptions and choices. Therefore, this highlights the need for experts to use an internationally standardised LCA approach to estimate CF of milk, for instance PAS 2050 or IDF (2010) dairy LCA guidelines, which can be externally certified by an accredited third party, e.g. Carbon Trust.

As expected, differences between grass-based farms' CF of milk were primarily explained by variation in measures of farm performance, particularly farm N efficiency. Therefore, this suggests grass-based farmers can reduce CF of milk and maintain or increase farm profit by adopting management practices that increase efficiency, e.g. improving cow total genetic merit. However, no individual farm attribute analysed explained the majority of variation between the CF of milk from grass-based farms. Therefore, the study indicates that several farm practices are required to reduce CF of milk from grass-based farms.

REFERENCES

1. Beukes PC, Gregorini P, Romera AJ, Levy G, Waghorn GC (2010) Improving production efficiency as a strategy to mitigate greenhouse gas emissions on pastoral dairy farms in New Zealand. Agric Ecosyst Environ 136(3–4):358–365

2. BSI (2011) PAS 2050:2011—specification for the assessment of life cycle green-house gas emissions of goods and services. British Standards Institute, London

3. Carbon Trust (2013) Carbon-footprinting software—footprint expert. The Carbon Trust, Dorset House, Stamford Street, London. http://www.carbontrust.com/software

4. Casey JW, Holden NM (2005) The relationship between greenhouse gas emissions and the intensity of milk production in Ireland. J Environ Qual 34(2):429–436

5. CSO (2013) Agriculture and fishing statistical products. Central Statistics Office, Skehard Road, Cork, Ireland. Accessed December 12, 2012. http://www.cso.ie/px/pxeirestat/statire/SelectTable/Omrade0.asp?Planguage=0

6. Christie KM, Rawnsley RP, Eckard RJ (2011) A whole farm systems analysis of greenhouse gas emissions of 60 Tasmanian dairy farms. Anim Feed Sci Technol 166–167:653–662

7. DAFM (2011) AIM bovine statistics report 2011. Department of Agriculture, Food and Marine, NBAS Division, Backweston, Co. Kildare

8. Dairy Australia (2012) Summary of the final report: farming—carbon footprint of the Australian dairy industry. Dairy Australia, Southbank

9. DairyCo (2012) Greenhouse gas emissions on British dairy farms, DairyCo carbon footprinting study: year one. Agriculture & Horticulture Development Board, Kenilworth

10. Defra (2007) The environmental, social and economic impacts associated with liquid milk consumption in the UK and its production: a review of literature and evidence. Department for Environment, Food and Rural Affairs, Nobel House, Smith square, London

11. Deighton M, O'Brien D, O'Loughlin B, O'Neill B, Wims C (2010) Towards reducing the methane intensity of milk production. TResearch 5(4):26–27

12. Duffy P, Hyde B, Hanley E, Dore C (2011a) Ireland informative inventory report 2011. Air pollutant emissions in Ireland 1990–2009 reported to the secretariat of the UN/ECE on long range transboundary air pollution. Environmental Protection Agency, Johnstown Castle Estate, Co. Wexford, Ireland

13. Duffy P, Hyde B, Hanley E, Dore C, O'Brien P, Cotter E, Black K (2011b) Ireland national inventory report 2011. Greenhouse gas emissions 1990–2009 reported to the United Nations framework convention on climate change. Environmental Protection Agency, Johnstown Castle Estate, Co. Wexford, Ireland

14. Ecoinvent (2010) Ecoinvent Centre. Ecoinvent 2.0 database. Swiss Centre for Life Cycle Inventories, Dübendorf. www.ecoinvent.ch

15. Evans RD, Dillon P, Shalloo L, Wallace M, Garrick DJ (2004) An economic comparison of dual-purpose and Holstein-Friesian cow breeds in a seasonal grass-based system under different milk production scenarios. Ir J Agric Food Res 43(1):1–16

16. Fitzgerald JJ, Mee JF, O'Grady D (2004) Systems of winter milk production based on all autumn calving cows. End of project report no. 4628. Dairy Production Department, Teagasc, Moorepark Research Centre, Fermoy, Co. Cork

17. Flysjö A, Cederberg C, Henriksson M, Ledgard S (2011) How does co-product handling affect the carbon footprint of milk? Case study of milk production in New Zealand and Sweden. Int J Life Cycle Assess 16(5):420–430

18. Gerber P, Vellinga T, Opio C, Henderson B, Steinfeld H (2010) Greenhouse gas emissions from the dairy sector. A life cycle assessment. Food and Agricultural Organization of the United Nations: Animal Production and Health Division, Viale delle Terme di Caracalla, 00153 Rome, Italy

19. Hagemann M, Ndambi A, Hemme T, Latacz-Lohmann U (2012) Contribution of milk production to global greenhouse gas emissions. Environ Sci Pollut Res 19(2):390–402

20. Howley M, Dennehy E, Holland M, O'Gallachoir B (2011) Energy in Ireland 1990–2010. Energy policy statistical support unit. Sustainable Energy Authority of Ireland

21. Hyde BP, Carton OT, O'Toole P, Misselbrook TH (2003) A new inventory of ammonia emissions from Irish agriculture. Atmos Environ 37(1):55–62

22. ICBF (2012) Dairy herd plus. Irish Cattle Breeding Federation, Shinagh, Bandon, Co. Cork, Ireland. http://www.icbf.com/?page_id=149

23. IDF (2010) A common carbon footprint for dairy. The IDF guide to standard life-cycle assessment methodology for the dairy industry. Bulletin of the International Dairy Federation 445, 38 pp

24. IPCC (2006) Intergovernmental Panel on Climate Change guidelines for national greenhouse inventories. In: Eggleston HS, Buendia L, Miwa K, Ngara T, Tanabe K (eds) Agriculture, forestry and other land use, vol 4, Institute for Global Environmental Strategies (IGES). Hayama, Japan

25. IPCC (2007) Changes in atmospheric constituents and in radiative forcing. In: climate change 2007: the physical science basis. Contribution of working group I to the fourth assessment report of the Intergovernmental Panel on Climate Change. Solomon S, Qin D, Manning M, Chen Z, Marquis M, Averyt KB, Tignor M., Miller HL (eds). Cambridge University Press, Cambridge

26. IPCC (2013) Anthropogenic and natural radiative forcing. In: climate change 2013: the physical science basis. Contribution of working group I to the fifth assessment report of the Intergovernmental Panel on Climate Change. Stocker TF, Qin D, Plattner GK, Tignor M, Allen SK, Boschung J, Nauels A, Xia Y, Bex V, Midgley PM (eds). Cambridge University Press, Cambridge

27. Jarrige R (1989) Ruminant nutrition: recommended allowances and feed tables. John Libbey Eurotext, Montrouge

28. Kennedy E, O'Donovan M, Murphy JP, Delaby L, O'Mara F (2005) Effects of grass pasture and concentrate-based feeding systems for spring-calving dairy cows in early spring on performance during lactation. Grass Forage Sci 60(3):310–318

29. LEAP (2014) Environmental performance of animal feed supply chains: guidelines for quantification. Livestock environmental assessment and performance partnership. FAO, Rome

30. Leip A, Weiss F, Wassenaar T, Perez I, Fellmann T, Loudjani P, Tubiello F, Grandgirard D, Monni S, Biala K (2010) Evaluation of the livestock's sector contribution to the EU greenhouse gas emissions (GGELS)—final report. European Commission, Joint Research Center, Ispra

31. Little (2002) Global comparative analysis of HFC and alternative technologies for refrigeration, air conditioning, foam, solvent, aerosol propellant, and fire protection applications, Final report to the alliance for responsible atmospheric policy. Arthur D. Little, Inc, Acorn Park, Cambridge

32. Lovett DK, Shalloo L, Dillon P, O'Mara FP (2008) Greenhouse gas emissions from pastoral based dairying systems: the effect of uncertainty and management change under two contrasting production systems. Livest Sci 116(1–3):260–274

33. Met Eireann (2013) Climate of Ireland—air temperature. http://www.met.ie/climate-ireland/surface-temperature.asp

34. O'Brien D, Shalloo L, Grainger C, Buckley F, Horan B, Wallace M (2010) The influence of strain of Holstein-Friesian cow and feeding system on greenhouse gas emissions from pastoral dairy farms. J Dairy Sci 93(7):3390–3402

35. O'Brien D, Shalloo L, Patton J, Buckley F, Grainger C, Wallace M (2012) A life cycle assessment of seasonal grass-based and confinement dairy farms. Agric Syst 107:33–46

36. O'Brien D, Shalloo L, Crosson P, Donnellan T, Farrelly N, Finnan J, Hanrahan K, Lalor S, Lanigan G, Thorne F, Schulte R (2014) An evaluation of the effect of greenhouse gas accounting methods on a marginal abatement cost curve for Irish agricultural greenhouse gas emissions. Environ Sci Pol 39:107–118

37. Olesen JE, Schelde K, Weiske A, Weisbjerg MR, Asman WAH, Djurhuus J (2006) Modelling greenhouse gas emissions from European conventional and organic dairy farms. Agric Ecosyst Environ 112(2–3):207–220

38. Olmos G, Mee JF, Hanlon A, Patton J, Murphy JJ, Boyle L (2009) Peripartum health and welfare of Holstein-Friesian cows in a confinement-TMR system compared to a pasture-based system. Anim Welf 18:467–476

39. O'Mara F (1996) A net energy system for cattle and sheep, Department of Animal Science and Production, Faculty of Agriculture. University College Dublin, Belfield

40. O'Mara F (2006) Development of emission factors for the Irish cattle herd, Environmental Protection Agency. Johnstown Castle, Co, Wexford

41. Opio C, Gerber P, Mottet A, Falcucci A, Tempio G, MacLeod M, Vellinga T, Henderson B, Steinfeld H (2013) Greenhouse gas emissions from ruminant supply chains—a global life cycle assessment. Food and Agriculture Organization of the United Nations (FAO), Rome

42. Ramsbottom G, Cromie AR, Horan B, Berry DP (2012) Relationship between dairy cow genetic merit and profit on commercial spring calving dairy farms. Animal 6(7):1031–1039

43. Robertson LJ, Waghorn GC (2002) Dairy industry perspectives of methane emissions and production from cattle fed pasture or total mixed rations in New Zealand. Proc N Z Soc Anim Prod 62:213–218

44. Rotz CA, Montes F, Chianese DS (2010) The carbon footprint of dairy production systems through partial life cycle assessment. J Dairy Sci 93(3):1266–1282

45. Roy P, Nei D, Orikasa T, Xu Q, Okadome H, Nakamura N, Shiina T (2009) A review of life cycle assessment (LCA) on some food products. J Food Eng 90(1):1–10

46. SAS (2008) SAS user guide version 9.1.3. Statistical Analysis Systems Institute Inc, Cary

47. Schils RLM, Olesen JE, del Prado A, Soussana JF (2007) A review of farm level modelling approaches for mitigating greenhouse gas emissions from ruminant livestock systems. Livest Sci 112:240–251

48. Schils RLM, Verhagen A, Aarts HFM, Šebek LBJ (2005) A farm level approach to define successful mitigation strategies for GHG emissions from ruminant livestock systems. Nutr Cycl Agroecosyst 71(2):163–175
49. Soussana JF, Allard V, Pilegaard K, Ambus P, Amman C, Campbell C, Ceschia E, Clifton-Brown J, Czobel S, Domingues R, Flechard C, Fuhrer J, Hensen A, Horvath L, Jones M, Kasper G, Martin C, Nagy Z, Neftel A, Raschi A, Baronti S, Rees RM, Skiba U, Stefani P, Manca G, Sutton M, Tuba Z, Valentini R (2007) Full accounting of the greenhouse gas (CO2, N2O, CH4) budget of nine European grassland sites. Agric Ecosyst Environ 121(1–2):121–134
50. Soussana JF, Tallec T, Blanfort V (2010) Mitigating the greenhouse gas balance of ruminant production systems through carbon sequestration in grasslands. Animal 4(3):334–350
51. Teagasc (2011) Sectoral road map: dairying. www.teagasc.ie/publications/2011/761/Roadmap_Dairy.pdf
52. Thoma G, Popp J, Shonnard D, Nutter D, Matlock M, Ulrich R, Kellogg W, Kim DS, Neiderman Z, Kemper N, Adom F, East C (2013) Regional analysis of greenhouse gas emissions from USA dairy farms: a cradle to farm-gate assessment of the American dairy industry circa 2008. Int Dairy J 31(Supplement 1):S29–S40
53. Thomassen MA, Dolman MA, van Calker KJ, de Boerd IJM (2009) Relating life cycle assessment indicators to gross value added for Dutch dairy farms. Ecol Econ 68(8–9):2278–2284
54. van Groenigen JW, Kuikman PJ, de Groot WJM, Velthof GL (2005) Nitrous oxide emission from urine-treated soil as influenced by urine composition and soil physical conditions. Soil Biol Biochem 37(3):463–473
55. Yan T, Agnew RE, Gordon FJ, Porter MG (2000) Prediction of methane energy output in dairy and beef cattle offered grass silage-based diets. Livest Prod Sci 64(2–3):253–263
56. Yan M-J, Humphreys J, Holden NM (2011) An evaluation of life cycle assessment of European milk production. J Environ Manag 92(3):372–379
57. Yan MJ, Humphreys J, Holden NM (2013a) The carbon footprint of pasture-based milk production: can white clover make a difference? J Dairy Sci 96(2):857–865
58. Yan MJ, Humphreys J, Holden NM (2013b) Life cycle assessment of milk production from commercial dairy farms: the influence of management tactics. J Dairy Sci 96(7):4112–4124

Evaluation of Industrial Dairy Waste (Milk Dust Powder) for Acetone-Butanol-Ethanol Production by Solventogenic *Clostridium* Species

VICTOR UJOR, ASHOK KUMAR BHARATHIDASAN, KATRINA CORNISH, AND THADDEUS CHUKWUEMEKA EZEJI

4.1 INTRODUCTION

Alternative energy-related research currently receives tremendous attention, largely in response to the rising cost of gasoline, and increased depletion of fossil fuel reserves. Consequently, interest in acetone-butanol-acetone (ABE) fermentation, which dwindled following the advent of the petrochemical industry, has been revived (Qureshi and Blaschek 2000; Qureshi and Maddox 2005; Yu et al. 2007). However, a major challenge hampering re-commercialization of the ABE process is lack of economic competitiveness, stemming in part from the absence of inexpensive, readily available, and easily fermentable substrates capable of generating high

Evaluation of Industrial Dairy Waste (Milk Dust Powder) for Acetone-Butanol-Ethanol Production by Solventogenic Clostridium *Species.* © *Ujor V, Bharathidasan AK, Cornish K, and Ezeji TC.* SpringerPlus *3,187 (2014), doi:10.1186/2193-1801-3-387. Licensed under a Creative Commons Attribution 4.0 International License, http://creativecommons.org/licenses/by/4.0/.*

ABE yields (Qureshi and Blaschek 2000; Yu et al. 2007). Interestingly, solventogenic *Clostridium* species are capable of fermenting a wide range of carbohydrates (Ezeji and Blaschek 2008), and lignocellulosic biomass has been identified as a potential substrate for inexpensive production of ABE and other fine chemicals (Ezeji and Blaschek 2008; Zhang and Ezeji 2013). However, bioconversion of lignocellulosic biomass is currently plagued by a number of limitations, notably generation of microbial inhibitory compounds during pretreatment and hydrolysis of lignocellulose to mixed sugars (Almeida et al. 2007), and inefficient utilization of the generated mixed sugars by fermenting microorganisms due to carbon catabolite repression (Ren et al. 2010). Therefore, given the broad substrate spectrum of solventogenic *Clostridium* species (Ezeji and Blaschek 2008; Servinsky et al. 2010; Yu et al. 2007), other cheap and readily utilizable substrates, whose applications in fermentation do not require pretreatment, may prove to be more cost-effective and efficient substrates than lignocellulose.

In comparison to lignocellulose, non-lignocellulosic substrates have generally been under-investigated for bio-butanol production. Some of the non-lignocellulosic substrates investigated, thus far, for ABE production include Jerusalem artichoke extract (Maddox 1980), cassava starch (Thang et al. 2010), cheese whey (Ennis and Maddox 1989; Maddox 1980; Qureshi and Maddox 2005; Stevens et al. 1988; Welsh and Veliky 1984), apple pomace (Voget et al. 1985) and starch-based packing peanuts (Jesse et al. 2002). Among these substrates, cheese whey is the most widely researched for ABE production, mostly due to its abundance, and high biological oxygen demand (BOD), which constitutes a major disposal predicament (Maddox 1980). Although some success has been recorded with whey, its use in large scale fermentation is plagued by a number of challenges, particularly its low sugar content (5%), which often warrants an initial concentration step prior to fermentation (Maddox 1980). Although glucose is the preferred sugar for solventogenic *Clostridium* species, different researchers have shown that lactose (the sugar content of whey) is utilized by these microorganisms when supplied as the sole carbon source (Bahl et al. 1986; Maddox 1980; Yu et al. 2007). Further, lactose metabolism favors butanol production over acetone (Bahl et al. 1986; Maddox

1980), an added economic incentive in light of the current impetus for butanol production.

Against a backdrop of the challenges associated with cheese whey fermentation to ABE, there is need to evaluate inexpensive non-whey-lactose-based substrates for ABE fermentation. Milk dust powder is such a substrate. Milk dust is a blend of different milk powders left over after industrial milk packaging. Milk dust powder constitutes a considerable hazard because suspensions of milk dust at 75–1000 g/m^3 of air can explode or self-ignite upon contact with hot surfaces (Davis et al. 2011; Ministry of Labor, New Zealand 1993). A drier with a capacity of 10 metric tons per h can generate 80.3 kg of airborne milk dust powder (Prevention of dust explosions in the food industry (2010)). The projected milk dust powder production (including skimmed and whole milk powder) in the USA in 2012 was approximately 1.02 million metric tons, which translates into 116.44 metric tons per h (United States Department of Agriculture (USDA) 2012). At present, some of the milk dust powder generated from the dairy industry is used as livestock feed; however, this enormous amount of waste (equivalent to 1.15×10^{14} Joules/year) has potential as a substrate for ABE fermentation. Milk dust powder constitutes predominantly of lactose, with small amounts of protein, fat and minerals. In addition, it retains all the natural properties of milk, such as color, flavor and solubility. Hence, upon mixing with water, milk dust powder resembles raw milk in appearance. In addition, milk dust does not contain urea, citric, lactic, and uric acids, found in low amounts in whey (Chatzipashali and Stamatis 2012; Foda et al. 2010; Napoli 2009), which can adversely affect fermentation efficiency.

Among solvent-producing *Clostridium* species, (including *C. saccharaoperbutylacetonicum, C. saccharobutylicum, C. tyrobutyricum,* and *C. pasterianum*), *Clostridium acetobutylicum* ATCC 824 and *Clostridium beijerinckii* NCIMB 8052 are the most characterized solventogenic *Clostridium* species and strains to date, and both have been used in the fermentation of whey, with varying outcomes (Ennis and Maddox 1989; Maddox 1980; Qureshi and Maddox 2005; Stevens et al. 1988; Welsh and Veliky 1984). However, most of these studies were centered on *C. acetobutylicum*, perhaps due to higher ABE concentrations obtained with this species relative to *C. beijerinckii*. Consequently, molecular characterization of lactose transport and utilization, and the underlying regulatory machineries in *C. acetobutylicum* ATCC 824 has been vigorously pursued (Servinsky et al. 2010; Yu et al. 2007). Albeit subject to full characterization, genomic information on *C. beijerinckii* NCIMB 8052 shows the presence of many (about 47) phosphoenol-pyruvate (PEP)-dependent phosphotransferase system (PTS) genes apparently involved in the metabolism of complex carbohydrates (Shi et al. 2010), including multiple genes whose protein products are putatively involved in lactose transport and metabolism. It is likely, therefore, that lactose metabolism in *C. beijerinckii* NCIMB 8052 is similar to *C. acetobutylicum* ATCC 824. To investigate this, we evaluated ABE production from milk dust powder by *C. beijerinckii* NCIMB 8052 and *C. acetobutylicum* ATCC 824.

4.2 MATERIALS AND METHODS

4.2.1 CHARACTERIZATION OF MILK DUST POWDER

Milk dust powder used in this study was obtained from International Dairy Ingredients, Inc. (Wapakoneta, Ohio, USA). Milk dust powder was recovered from the dust collector following spray-drying of milk. Prior to fermentation, milk dust powder was subjected to a series of analyses to determine the nitrogen, ash, mineral, energy (calorific value), total solid and moisture, and total organic carbon contents (methods details below).

4.2.2 DETERMINATION OF ASH CONTENT

The ash content of milk dust powder was analyzed according to the procedure described in the Test Methods for the Examination of Composting and Compost (TMECC 2002). Pre-weighed samples were ignited in a forced air muffle furnace (Barnstead Thermolyne 30400-Series Furnace; Model: F30428C-80) in the presence of excess air at 550°C for 2–3 h followed by cooling in a desiccator at room temperature. The resulting ash was weighed and estimated as the percentage of ash content (dry wt., w/w) in each sample.

4.2.3 DETERMINATION OF TOTAL SOLID AND MOISTURE CONTENT

Total solids and moisture content were assessed using a modified version of the TMECC method 03.09-A (2002). To prevent sugar caramelization, samples of milk dust powder were subjected to drying at $50 \pm 5°C$, as opposed to $70 \pm 5°C$ recommended in the TMECC protocol (TMECC method 03.09-A 2002). Drying was allowed to proceed for 24 h or until there was no further detectable change in weight. Total solid was reported as percentage of dry solid contained in the fresh sample.

4.2.4 ESTIMATION OF CALORIFIC CONTENT (ENERGY VALUE)

Calorific value is a measure of the fat, carbohydrate and protein content of a food material. The calorific content of milk dust powder was determined with a Bomb (combustion) calorimeter (Model: C 2000 Basic version 1, IKA). Samples (200 mg) were introduced into the decomposition unit (Model: C 5010) of the Bomb calorimeter, and incinerated in the presence of pure oxygen. The Bomb calorimeter estimates gross calorific value as the quotient of the amount of heat liberated upon total combustion and the weight of the original sample. Calorific contents were determined for triplicate samples.

4.2.5 MEASUREMENT OF TOTAL ORGANIC CARBON (TOC)

Total organic carbon (TOC) includes biodegradable sugars, protein, and fat content of a material but does not include inorganic carbonate fractions such as calcium and magnesium carbonates. The TOC of milk dust powder was measured in accordance with the TMECC method 04.01-A (2002), using a carbon analyzer (Model Vario MAX CN, Elementar Americas).

Samples were briefly subjected to combustion in an oxygen-rich atmosphere in a resistance furnace at 1,370°C. The CO_2 produced was passed through an oxygen stream in anhydrone tubes to scrub water vapor out of the stream. The dehydrated CO_2 stream was then channeled into an infrared detector, which generates a signal proportionate to the amount of CO_2 detected. The resulting values are reported as percentage of TOC content (dry wt., w/w) in dried samples.

4.2.6 QUANTIFICATION OF TOTAL NITROGEN (TN) CONTENT

Total nitrogen is the sum of organic- and ammonia-derived nitrogen (nitrogen from proteins, nitrates and nitrites). The total nitrogen content of a material facilitates the determination of its carbon to nitrogen ratio (C:N). C:N ratio helps to determine the fermentability of a substrate, because nitrogen is largely essential for cell growth, while the carbon content of a substrate is critical to product yield, in this case ABE. The total nitrogen content of milk dust was determined according to the TMECC 04.02-D method (oxidation by dry combustion), by employing an automated Nitrogen analyzer (Model: Vario MAX CN, Elementar Americas). Samples (150 mg) were combusted in an oxygen-rich chamber, at a temperature of about 900°C to generate a gas stream containing CO_2, H_2O, and N_2. The gas stream then was fed into a separation column, which specifically removes CO_2 and H_2O. The pure N_2 was passed into a thermal conductivity detector, which generates a signal proportional to the amount of N_2 produced. Total nitrogen is presented as percentage content (dry wt., w/w) of dried samples.

4.2.7 ANALYSIS OF ELEMENTAL COMPOSITION OF MILK DUST POWDER

Elemental composition of milk dust was measured using an inductively coupled plasma optical emission spectroscope (ICP-OES, Teledyne Leeman Labs Prodigy) following protocols described in the TMECC methods (sections 04.05, 04.06 and 04.07; 2002). Fully dried samples (1 g) were transferred into polytetrafluroethylene (PTFE) or Teflon vessels and solubilized in concentrated HNO_3 (7 mL). Samples were microwave-digested as described in the TMECC method (section 04.12-A). When compared to alternative methods, microwave digestion allows for a more rapid digestion as it employs high pressure and temperature within the vessels and the use of closed vessel prevents cross-contamination among samples and loss of volatile elements (Sun et al. 2000). The digested samples were allowed to cool to room temperature before being transferred to an ICP-OES auto-sampler for analysis. The ICP uses argon (~10–15 L/min) to ionize the digested samples in an applied radio frequency field. Post ionization, each element exhibits a distinctive emission spectrum, of which the identity and intensity is detected and quantified by the detector (Sun et al. 2000). The concentration of each element is expressed as a function of the intensity of the corresponding elemental spectrum. The resulting concentrations are reported in mg/g sample (on a dry weight basis).

4.2.8 BACTERIAL STRAINS AND CULTURE CONDITIONS

C. beijerinckii NCIMB 8052 (C. beijerinckii ATCC 51743; hereafter referred to as C. beijerinckii) and C. acetobutylicum ATCC 824 (hereafter referred to as C. acetobutylicum) obtained from the American Type Culture Collection (Manassas, VA) were used in the fermentation of milk dust powder. Spores were stored in sterile, double-distilled water at 4°C. To revive spores for inoculation, stocks (200 µL) were heat-shocked for 10 min at 75°C followed by cooling on ice. The heat-shocked spores were then inoculated into anoxic pre-sterilized tryptone–glucose–yeast extract (TGY) broth (10 ml) and incubated in an anaerobic chamber (Coy Laboratory Products Inc., Ann Arbor, Michigan), with a modified atmosphere of 82% N_2, 15% CO_2, and 3% H_2 for 12 h to 14 h at 35°C ± 1°C. When the optical density (OD600nm) reached 0.9–1.1, 8 ml of actively growing culture was transferred into 92 mL of anoxic TGY medium and incubated as above until the OD600nm reached of 0.9–1.1 (Zhang and Ezeji 2013; Han et al. 2011). This was used as the pre-culture to inoculate the milk dust-based fermentations.

Batch ABE fermentation by *C. beijerinckii* and *C. acetobutylicum* was performed in 150 ml Pyrex screw-capped media bottles containing anoxic milk dust-based medium (in a final volume of 100). The milk dust powder medium was prepared by autoclaving a mixture of ~12 g milk dust powder and 1g yeast extract at 121°C for 15 min. Upon cooling to 40°C, the mixture was transferred into the anaerobic chamber and ~80 ml anoxic, sterilized distilled water was added to bring the final concentration of lactose in the medium to 50 g/L. Prior to inoculation, 1ml each of filter-sterilized P2 stocks including vitamin (0.1 g/L para-amino-benzoic acid; 0.1 g/L thiamine; 0.001 g/L biotin), buffer (50 g/L KH_2PO_4; 50 g/L K_2HPO_4; 220 g/L ammonium acetate) and mineral (20 g/L $MgSO_4.7H_2O$; 1 g/L $MnSO_4$. H_2O; 1 g/L $FeSO_4.7H_2O$; 1 g/L NaCl) solutions were added (Richmond et al. 2012; Zhang et al. 2012). As a control, 100 ml of P2 medium (glucose, 60 g/L and yeast extract, 1 g/L) containing P2 stock solutions was inoculated with both *C. acetobutylicum* and *C. beijerinckii*. All cultures were inoculated with 6% pre-culture for both species studied. Samples were taken every 12 h for pH, residual sugars, ABE, and acid analyses. Unless otherwise stated, all fermentations were conducted in triplicate at $35 \pm 1°C$, and no agitation or pH control was employed.

4.2.9 ANALYTICAL PROCEDURES

Owing to the cloudiness of the milk dust medium, bacterial growth was determined by plate count (viable cell count). Each colony forming unit (CFU) was regarded to have originated from a single cell. Culture samples were serially diluted in 10 ml of TGY medium and 100 μL of serially diluted samples were plated on 10 ml semi-solid TGY agar (0.45% agar in TGY medium). The plates were incubated anaerobically for 24–48 h at $35 \pm 1°C$ and the number of CFUs was counted and expressed as CFU per ml of original culture (Jesse et al. 2002; Nielsen et al. 2009). Concentrations of glucose and lactose were measured by high performance liquid chromatography (HPLC) with a refractive index (RI) detector (Agilent Technologies 1200 Series) using an organic acid column (Rezex ROA-Organic Acid H^+ column, 300 mm × 7.8 mm). The mobile phase was 0.0025 M H_2SO_4 (Fluka) operated at a flow rate of 0.6 ml/min. All samples were injected by automatic sampler and the injection volume was 10 μL. The column and detector temperature were maintained at 80°C and 55°C respectively.

The pH profiles of fermentation cultures were monitored with a Beckman Φ500 pH meter (Beckman Coulter Inc., Brea, CA). Concentrations of fermentation products, namely, acetate, butyrate, acetone, butanol, and ethanol, were measured using a 7890A Agilent Technologies gas chromatograph (Agilent Technologies Inc., Wilmington, DE) equipped with a flame ionization detector (FID) and 30 m (length) × 320 μm (internal diameter) × 0.50 μm (HP-Innowax film) J x W 19091N-213 capillary column as described previously (Han et al. 2011, 2012), with nitrogen as carrier gas. The inlet and detector temperatures were maintained at 250°C and 300°C respectively. The temperature of the oven was programmed from 60–200°C with 20°C/min increments, and a 5-min hold at 200°C. One microliter was injected per sample with a split ratio of 10:1. Yield was calculated as the maximum amount of butanol/ABE produced per gram of substrate utilized (expressed in g/g substrate). ABE productivity was estimated as maximum ABE produced (g/L) divided by the corresponding fermentation time in hours (Ezeji and Blaschek 2008).

4.3 RESULTS

4.3.1 THE PHYSICO-CHEMICAL CHARACTERISTICS AND COMPOSITION OF MILK DUST POWDER

HPLC analysis showed that the milk dust powder used in this study had lactose content of ~425 g/L (w/v; wet wt.); 8.5-fold higher than cheese whey, which typically has a lactose content of about 50 g/L (Welsh and Veliky 1984). The total solid and calorific contents, and the carbon:nitrogen ratio of the milk dust powder were ~94%, 50 kJ/kg and 7, respectively (Table 1). The predominating elements were potassium (K), phosphorus (P), and calcium (Ca) with concentrations of 13.05, 6.38, and 7.08 mg/g dry matter, respectively (Table 1). Copper (Cu) was the least element present with a concentration of 0.0004 mg/g dry matter (Table 1).

TABLE 1: The physico chemical properties of milk dust powder

% Ash	% Total solids	Calorific value (kJ/kg)	% Carbon	% Nitrogen	C/N ratio
3.79 ±0.21	93.64 ±0.10	50.20 ±1.92	45.51 ±0.30	6.46 ±0.10	7.04 ±0.33

Elemental composition (mg/g dry matter)

P	K	Ca	Mg	S	Al	B	Cu	Fe	Mn	Mo	Na	Zn
6.38 ±0.15	13.05 ±0.31	7.08 ±0.25	0.89 ±0.15	4.02 ±0.33	0.03 ±0.11	0.01 ±0.12	0.0004 ±0.22	0.03 ±0.36	0.0002 ±0.08	0.001 ±0.05	4.03 ±0.38	0.03 ±0.05

TABLE 2: Growth (colony-forming units/ml) of *C. acetobutylicum* and *C. beijerinckii* on milk dust powder based medium

Time (h)	Glucose*C. acetobutylicum*ATCC 824 cfu/ml	Milk dust*C. acetobutylicum*ATCC 824 cfu/ml	Glucose*C. beijerinckii*NCIMB 8052 cfu/ml	Milk dust*C. beijerinckii*NCIMB 8052 cfu/ml
0	2.86×10^7	1.87×10^7	5.1×10^7	1.17×10^7
12	8.06×10^9	3.1×10^8	2.0×10^9	1.27×10^8
24	4.46×10^{10}	4.0×10^9	3.3×10^9	4.67×10^8
36	1.39×10^{11}	7.67×10^9	1.1×10^{10}	3.67×10^9
48	8.3×10^{10}	1.03×10^9	8.6×10^9	2.33×10^9
60	6.1×10^{10}	1.13×10^9	6×10^9	1.3×10^8
72	6.9×10^9	7.0×10^8	1.2×10^9	1.6×10^8

4.3.2 GROWTH PROFILES OF C. ACETOBUTYLICUM AND C. BEIJERINCKII ON MILK DUST MEDIUM

To assess the growth of *C. acetobutylicum* and *C. beijerinckii* on milk dust medium, the growth profiles of both species on milk dust medium were compared against each other and in relation to control fermentations on glucose, their preferred substrate. Microbial cell counts showed that both *C. acetobutylicum* and *C. beijerinckii* grew better on glucose than on lactose (milk dust) (Table 2). Maximum cell count obtained with glucose-grown cultures of *C. acetobutylicum* (1.39×10^{11}) was 18-fold higher than the maximum count for cultures grown on milk dust medium (7.67×10^9). The growth of *C. beijerinckii* on glucose (1.1×10^{10}) was only 3-fold better than on milk dust medium (3.67×10^9). Compared to *C. beijerinckii*, the colony-forming units of *C. acetobutylicum* were 2.1-fold higher, when both microorganisms were grown on milk dust powder. It is noteworthy, however, that the improved growth of *C. acetobutylicum* over *C. beijerinckii* was more pronounced when both were grown on glucose, in which the cell count for *C. acetobutylicum* was ~13-fold higher than that obtained with *C. beijerinckii*.

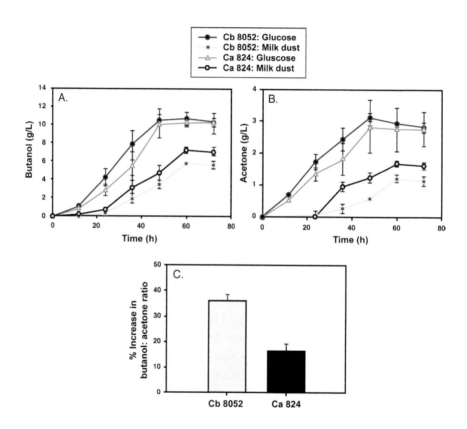

FIGURE 1: The concentrations of butanol , and acetone of *C. acetobutylicum* and *C. beijerinckii* on milk dust powder, and their butanol:acetone ratios relative to glucose grown cultures. (A) Butanol produced during ABE fermentation, (B) acetone produced during ABE fermentation, and (C) increase in butanol:acetone ratio in cultures grown in milk dust powder medium relative to cultures grown in glucose medium. Ca 824: *C. acetobutylicum*; Cb 8052: *C. beijerinckii*.

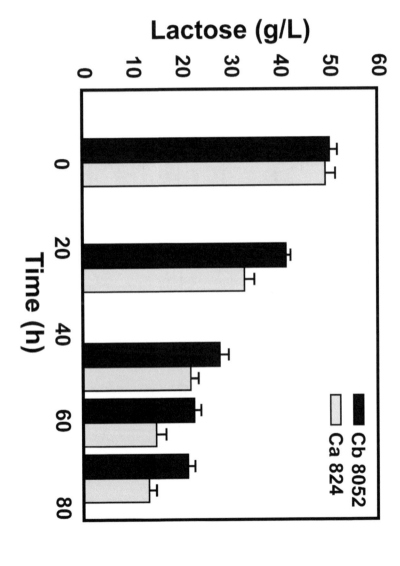

FIGURE 2: Lactose utilization profiles of *C. acetobutylicum* and *C. beijerinckii*. Ca 824: *C. acetobutylicum*; Cb 8052: *C. beijerinckii*.

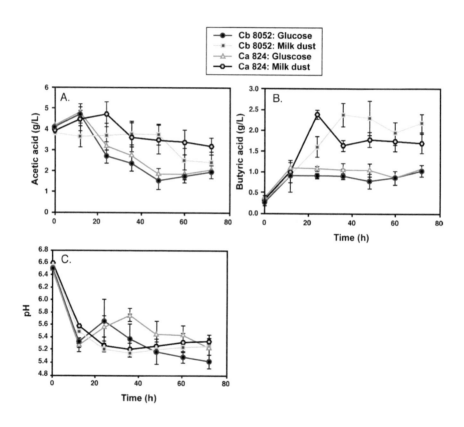

FIGURE 3: The levels of acetate and butyrate and pH profiles of cultures of *C. acetobutylicum* and *C. beijerinckii* grown on milk powder medium. (A) Acetic acid concentration profile during ABE fermentation, (B) butyric acid concentration profile during ABE fermentation, and (C) pH profile during ABE fermentation. Ca 824: *C. acetobutylicum*; Cb 8052: *C. beijerinckii*.

TABLE 3: Sugar utilization, ABE concentrations, yields and productivities of *C. acetobutylicum* and *C. beijerinckii*

Parameters	Glucose*C. acetobutylicum*ATCC 824	Milk dust*C. acetobutylicum*ATCC 824	Glucose*C. beijerinckii*NCIMB 8052	Milk dust*C. beijerinckii*NCIMB 8052
Acetone (g/L)	2.84 ± 0.82	1.70 ± 0.07	3.12 ± 0.12	1.21 ± 0.11
Ethanol (g/L)	1.39 ± 0.33	1.45 ± 0.21	1.33 ± 0.48	1.16 ± 0.21
Butanol (g/L)	**10.65 ± 0.69**	**7.25 ± 0.32**	**10.51 ± 0.65**	**5.80 ± 0.12**
Total ABE (g/L)	**14.62 ± 1.74**	**10.25 ± 0.62**	**14.97 ± 1.25**	**8.15 ± 0.79**
Initial glucose/ lactose (g/L)	60.89 ± 0.16	49.94 ± 1.61	60.86 ± 0.37	49.08 ± 1.26
Final glucose/ lactose (g/L)	**20.44 ± 1.92**	**14.36 ± 1.66**	20.24 ± 2.44	**22.26 ± 1.43**
Total glucose/ lactose utilized (g/L)	40.45 ± 1.77	34.72 ± 1.64	40.62 ± 2.08	27.67 ± 0.70
ABE yield (g/g of substrate)	0.36 ± 0.01	0.30 ± 0.03	0.37 ± 0.02	0.29 ± 0.02
ABE productivity (g/L/h)	0.25 ± 0.01	0.17 ± 0.01	0.31 ± 0.02	0.13 ± 0.01

4.3.3 FERMENTATION PROFILES OF C. ACETOBUTYLICUM AND C. BEIJERINCKII

As with growth, both microorganisms performed better on glucose than in milk dust-based medium. *C. acetobutylicum* produced more ABE than *C. beijerinckii* when both were grown in milk dust medium (Table 3, Figure 1). Cultures of *C. acetobutylicum* grown in milk dust medium produced 7.3 and 10.3 g/L of butanol and ABE, respectively. Both were 1.3-fold higher than the butanol (5.8 g/L) and ABE (8.2 g/L) concentrations produced by *C. beijerinckii* in milk dust medium (Table 3, Figure 1A). However, a preponderance of butanol over acetone in cultures grown on lactose as the main carbon source (Maddox 1980) was observed with both species. Whereas *C. acetobutylicum* produced 1.7 g/L acetone, *C. beijerinckii* produced 1.2 g/L (Figure 1B) when grown in milk dust medium. As a result, butanol:acetone ratios for both microorganisms were higher on milk dust relative to glucose. In fact, the ratios of butanol to acetone in cultures of *C.*

beijerinckii and *C. acetobutylicum* grown on milk dust powder were 36% and 16% higher, respectively, relative to glucose-grown cultures (Figure 1B). This is over 2-fold higher with *C. beijerinckii* when compared to *C. acetobutylicum*. The ABE yield of *C. acetobutylicum* did not vary significantly from that of *C. beijerinckii* despite producing a higher ABE concentration (Table 3). This is ascribable to higher lactose consumption by *C. acetobutylicum*, which utilized 34.7 g/L of lactose; 30% higher than the 27.7 g/L consumed by *C. beijerinckii* (Figure 2).

Notably, acetate and butyrate levels in cultures grown in milk dust medium were considerably higher than the levels detected in cultures grown on glucose (Figure 3). Both fermentation media (glucose- and milk dust-based) contain acetate (2.2 g/L) in addition to acetate carried over from the preculture. As a result, acetate levels at 0 h were ~4 g/L. In glucose-based medium, acetate levels dropped sharply, while they either increased or remained relatively stable in cultures grown on milk dust. Consequently, over the course of fermentation, acetate levels in milk dust media were 1.2- to 2.5-fold higher in cultures grown on milk dust when compared to those grown on glucose, over the course of fermentation (Figure 3). Similarly, butyrate levels were considerably higher in milk dust medium than in the glucose medium (1.3- to 2.5-fold) for both *C. acetobutylicum* and *C. beijerinckii*. Despite the differences in acetate and butyrate levels between cultures grown on glucose and those grown on lactose (milk powder-based medium), the pH profiles of both sets of cultures varied only slightly (Figure 3B). Variations in pH were more obvious with *C. acetobutylicum* than with *C. beijerinckii*. With *C. beijerinckii*, the pH values were slightly higher in the glucose medium than the milk dust powder medium only at 24 and 36 h. Conversely, with *C. acetobutylicum*, pH remained marginally higher in the glucose-grown cultures than those grown on milk dust from 24 to 60 h.

4.4 DISCUSSION

Return of ABE fermentation to industrial relevance is contingent upon the availability of readily fermentable inexpensive substrate(s), with high product yield (Yu et al. 2007). Our results demonstrate that milk dust powder can be fermented to butanol without the usual concentration, pH adjustment, and deproteination processes (Welsh and Veliky 1984) often associated with cheese whey fermentation. Further, our results substantiate previous reports (Bahl et al. 1986; Maddox 1980) that lactose fermentation favors butanol production over acetone, as both *C. acetobutylicum* and *C. beijerinckii*; particularly the latter showed higher butanol:acetone ratios on milk dust powder relative to glucose.

When compared to the glucose medium, a number of factors may account for the lower product concentrations and yields on milk dust medium (Table 2). First, the preference of glucose over lactose by solventogenic clostridia may be a dominant factor underlying this pattern. Secondly, the physico-chemical properties of milk dust powder may have negatively influenced growth and ABE production, as cultures grown on milk dust-based medium exhibited a prolonged lag phase, resulting in delayed ABE production (Figure 1). Whereas accumulation of ABE increased significantly 12 h after inoculation in the glucose medium, a substantial increase in ABE production was not observed in the milk powder medium until 24 h post inoculation (Figure 1). After heat-sterilization of milk powder, reconstitution in water resulted in the formation of hard, crystal-like balls, which reduced the medium surface area, thereby limiting sugar uptake. Milk dust powder constitutes largely of casein (~80%) and whey (20%) proteins (Farell et al. 2004). When milk dust is subjected to elevated temperatures (90°C–120°C for 10 or more min), casein and whey undergo coagulation and denaturation, respectively (Cleaning and Sanitizing of Containers and Equipment (2005); Coagulation Of Milk, Part 3 (2007); Bender 2005). This is because the stability of casein in milk dust during heat treatment is dependent on the availability of optimal amounts of calcium and magnesium (present in phosphate and citrate forms) which limit the alignment of casein into a three-dimensional lattice that holds sugars, fats, and water in position thus preventing their availability for microbial uptake. Although the milk dust powder used in this study was found to contain significant levels of calcium, magnesium and phosphorus (Table 1), heat treatment affects salt equilibria leading to the precipitation of calcium and magnesium phosphates/citrates (http://drinc.ucdavis.edu/dairyp/dairyp5.htm; http://www.chestofbooks.com/ food/ science/Experimental-Cookery/Coagulation-Of-Milk-Part-3.html; Bender 2005). Consequently, this aggravates casein coagulation, and the resulting crystals drastically reduce the availability of sugars to the fermenting cells. Physically breaking these hard crystals were found to enhance growth and ABE production (Figure 4C). In addition to influencing lactose utilization and ABE production, it is likely that these physical characteristics affected analytical measurement of lactose in solution given that the milk dust powder did not dissolve completely in water.

Thirdly, ABE fermentation is pH-sensitive because of the delicate interplay between fermenting cells and acids (acetate and butyrate), relative to ABE production (Grupe and Gottschalk 1992). However, we have observed that pH values within 5.2 and 5.8 favor ABE productions. Although the pH ranges of both microorganisms investigated in this study were well within this range during growth in milk dust medium, ABE production was nonetheless less than the levels observed with glucose. In view of this, we contend that the pH values might have been influenced by the heterogeneous nature of the milk dust powder medium (Figure 4), as the pH values do not mirror the corresponding acid levels. With the cloudy consistency of milk dust powder coupled with the presence of large particles, it is likely that pockets of microbial activity and inactivity or low activity may be distributed in the culture. In addition, the pH of ABE fermentation broth is significantly influenced by the ratio of protonated acetic and butyric acids to their unprotonated forms (Farell et al. 2004; Bryant and Blaschek 1988; Russell and Diez-Gonzalez 1998; Ezeji et al. 2010). Given the higher acid levels detected in cultures of *C. acetobutylicum* and *C. beijerinckii* grown on milk dust powder medium, relative to cultures grown on glucose, it may be deduced that acid reassimilation was impaired in the milk dust powder medium (Figure 3A and B). Poor acid reassimilation has been reported for substrate-limited cultures of solventogenic clostridia (Jesse et al. 2002). Reduced availability of lactose in the milk dust medium owing to coagulation of casein post sterilization is therefore a probable factor contributing to the higher acetic and butyric acid levels detected in this medium. Although reduced nutrient uptake should affect acid production pre-solventogenesis, the produced acids are less readily reabsorbed with nutrient limitation (Jesse et al. 2002), and the media used in this study (both glucose- and milk dust powder-based) contained 2.2 g/L acetic acid from the onset of fermentation. Taken together, these factors would ultimately dampen ABE production.

FIGURE 4: Physical properties of milk dust medium used for ABE fermentation by *C. acetobutylicum* and *C. beijerinckii*. (A) Coagulated milk dust powder prior to autoclaving, (B) coagulated milk dust powder after autoclaving, and (C) coagulated milk dust powder medium was shaken prior to inoculation with *C. acetobutylicum* or *C. beijerinckii*.

Clearly, *C. acetobutylicum* performed better than *C. beijerinckii* on milk dust medium with respect to growth (Table 2), lactose utilization (Figure 2) and ABE production (Table 3; Figure 1), but not on glucose. This may be indicative of superior lactose transport and metabolism by *C. acetobutylicum* relative *C. beijerinckii*. *C. acetobutylicum* has been shown to possess robust lactose metabolic machinery (Servinsky et al. 2010). On the other hand, there is a paucity of experimental data on the utilization of lactose by *C. beijerinckii*. Hence, studies targeted at unraveling the regulatory machineries that govern lactose utilization in *C. beijerinckii* may prove instructive. We anticipate that the findings of this study would encourage such investigation with a view to delineating the discrepancy in lactose utilization, and consequently, ABE production between *C. acetobutylicum* and *C. beijerinckii* on a lactose-replete medium.

This study demonstrates for the first time, fermentation of non-whey, lactose-rich industrial diary waste for ABE production. Fermentation of milk dust powder by both species studied favors butanol production over acetone, similar to previous reports for whey. Taken together, availability of cheap milk dust powder, calls for further investigation towards improving its fermentability for enhanced ABE production.

REFERENCES

1. Almeida JRM, Modig T, Petersson A, Hähn-Hägerdal B, Lidén G, Gorwa-Grauslund MF (2007) Increased tolerance and conversion of inhibitors in lignocellulosic hydrolysates by Saccharomyces cerevisiae. J Chem Technol Biotechnol 82:340-349
2. Bahl H, Gottwald M, Kuhn A, Rale V, Andersch W, Gottschalk G (1986) Nutritional factors affecting the ratio of solvent produced by *Clostridium* acetobutylicum. Appl Environ Microbiol 52:169-172
3. Bender DA (2005) "Milk-stone". A Dictionary of Food and Nutrition. http://www.encyclopedia.com/doc/1O39-milkstone.html; Accessed Nov 3, 2012
4. Bryant DL, Blaschek HP (1988) Buffering as a means for increasing the growth and butanol production of *Clostridium* acetobutylicum. J Ind Microbiol 3:49-55
5. Chatzipashali AA, Stamatis AG (2012) Biotechnological utilization with focus on anaerobic treatment of cheese whey: current status and prospects. Energies 5:3492-3525
6. Cleaning and Sanitizing of Containers and Equipment (2005) http://drinc.ucdavis.edu/dairyp/dairyp5.htm; Accessed Nov 3, 2012
7. Coagulation Of Milk, Part 3 (2007) http://www.chestofbooks.com/food/science/Experimental-Cookery/Coagulation-Of-Milk-Part-3.html; Accessed Nov 3, 2012

8. Davis SG, Hinze PC, Hansen OR, van Wingerden K (2011) Does your facility have a dust problem: methods for evaluating dust explosion hazards. J Loss Prev Process Ind 24:837-846

9. Ennis BM, Maddox IS (1989) Production of solvents (ABE fermentation) from whey permeate by continuous fermentation in a membrane reactor. Bioprocess Eng 4:27-34

10. Ezeji TC, Blaschek H (2008) Fermentation of dried distillers' soluble (DDGS) hydrolysates to solvents and value-added products by solventogenic Clostridia. Bioresour Technol 99:5232-5242

11. Ezeji T, Milne C, Price ND, Blaschek HP (2010) Achievements and perspectives to overcome the poor solvent resistance in acetone and butanol-producing microorganisms. Appl Microbiol Biotechnol 85:1697-1712

12. Farell H Jr, Jimenez-Flores R, Bleck G, Brown E, Butler J, Creamer L, Swaisgood H (2004) Nomenclature of the proteins of cows' milk–sixth revision. J Dairy Sci 87:1641-1674

13. Foda MI, Dong H, Li Y (2010) Study of the suitability of cheese whey for Biobutanol production by clostridia. J Am Sci 6:39-46

14. Grupe H, Gottschalk G (1992) Physiological events in *Clostridium* acetobutylicum during the shift from acidogenesis to solventogenesis in continuous cultures and presentation of a model for shift induction. Appl Environ Microbiol 58:3896-3902

15. Han B, Gopalan V, Ezeji TC (2011) Acetone production in solventogenic *Clostridium* species: New insights from non-enzymatic decarboxylation of acetoacetate. Appl Microbiol Biotechnol 91:565-576

16. Han B, Ujor V, Lai LB, Gopalan V, Ezeji TC (2012) Use of proteomic analysis to elucidate the role of calcium in acetone-butanol-ethanol (ABE) fermentation in *Clostridium* beijerinckii NCIMB 8052. Appl Environ Microbiol 79:282-293

17. Jesse TW, Ezeji TC, Qureshi N, Blaschek HP (2002) Production of butanol from starch-based waste packing peanuts and agricultural waste. J Ind Microbiol Biotechnol 29:117-123

18. Maddox IS (1980) Production of n-butanol from whey filtrate using *Clostridium* acetobutylicum NCIB 2951. Biotechnol Lett 2:493-498

19. Ministry of Labor, New Zealand (1993) Approved code of practice for the prevention, detection and control of fire and explosion in New Zealand dairy industry spray drying plant. 10-21

20. Napoli F (2009) Development of an integrated bioprocess for butanol production. Doctoral dissertation, Università Federico II.

21. Nielsen DR, Leonard E, Yoon SH, Tseng HC, Yuan C, Prather KLJ (2009) Engineering alternative butanol production platforms in heterologous bacteria. Metab Eng 11:262-273

22. Prevention of dust explosions in the food industry (2010) http://www.hse.gov.uk/food/dustexplosion.htm; Accessed Nov 3, 2012

23. Qureshi N, Blaschek H (2000) Butanol production using *Clostridium* beijerinckii BA101 hyper-butanol producing mutant strain and recovery by pervaporation. Appl Biochem Biotechnol 84:225-235

24. Qureshi N, Maddox I (2005) Reduction in butanol inhibition by perstraction: utilization of concentrated lactose/whey permeate by *Clostridium* acetobutylicum to enhance butanol fermentation economics. Food Bioprod Process 83:43-52

25. Ren C, Gu Y, Hu S, Wu Y, Wang P, Yang Y, Yang C, Yang S, Jiang W (2010) Identification and inactivation of pleitropic regulator CcpA to eliminate glucose repression of xylose utilization in *Clostridium* acetobutylicum. Metab Eng 12:446-454

26. Richmond C, Ujor V, Ezeji TC (2012) Impact of syringaldehyde on the growth of *Clostridium* beijerinckii NCIMB 8052 and butanol production. 3 Biotech 2:159-167

27. Russell JB, Diez-Gonzalez F (1998) The effects of fermentation acids on bacterial growth. Adv Microbial Physiol 39:205-234

28. Servinsky MD, Kiel JT, Dupuy NF, Sund CJ (2010) Transcriptional analysis of differential carbohydrate utilization by *Clostridium* acetobutylicum. Microbiology 156:3478-3491

29. Shi Y, Li YX, Li YY (2010) Large number of phosphotransferase genes in *Clostridium* beijerinckii NCIMB 8052 genome and study on their evolution. BMC Bioinformatics 11(Suppl 11):S9 doi:10.1186/1471-2105-11-S11-S9

30. Stevens D, Alam S, Bajpai R (1988) Fermentation of cheese whey by a mixed culture of *Clostridium* beijerinckii and Bacillus cereus. J Ind Microbiol 3:15-19

31. Sun DH, Waters JK, Mawhinney TP (2000) Determination of thirteen common elements in food samples by inductive coupled plasma atomic emission spectrometry: comparison of five digestion methods. J AOAC Int 83:1218-1224

32. Thang VH, Kanda K, Kobayashi G (2010) Production of acetone-butanol-ethanol (ABE) in direct fermentation of cassava by *Clostridium* saccharoperbutylacetonicum N1-4. Appl Biochem Biotechnol 161:157-170

33. TMECC (2002) Test methods for the examination of composting and compost (TMECC) USDA, and US composting council. http://compostingcouncil.org/admin/wp-content/plugins/wppdfupload/pdf/34/TMECC%20Purpose,%20Composting%20Process.pdf; Accessed November 3, 2012

34. United States Department of Agriculture (USDA) (2012) World markets and trade, foreign agricultural service report.

35. Voget C, Mignone C, Ertola R (1985) Butanol production from apple pomace. Biotechnol Lett 7:43-46

36. Welsh FW, Veliky IA (1984) Production of acetone-butanol from acid whey. Biotechnol Lett 6:61-64

37. Yu Y, Tangey M, Aass HC, Mitchel WJ (2007) Analysis of the mechanism and regulation of lactose transport and metabolism in *Clostridium* acetobutylicum ATCC 824. Appl Environ Microbiol 73:1842-1850

38. Zhang Y, Ezeji TC (2013) Transcriptional analysis of *Clostridium* beijerinckii NCIMB 8052 to elucidate role of furfural stress during acetone butanol ethanol fermentation. Biotechnol Biofuels 6:66

39. Zhang Y, Han B, Ezeji TC (2012) Biotransformation of furfural and 5-hyroxymethylfurfural (HMF) by *Clostridium* acetobutylicum ATCC 824 during butanol fermentation. New Biotechnol 29:345-351

CHAPTER 5

Life Cycle Assessment of Cheese and Whey Production in the USA

DAESOO KIM, GREG THOMA, DARIN NUTTER, FRANCO MILANI, RICK ULRICH, AND GREG NORRIS

5.1 INTRODUCTION

Consumers are increasingly aware of the sustainability characteristics of the products they purchase. As a result, a key issue for the dairy industry is ensuring that dairy manufacturing, especially cheese in this study, is conducted with sustainability in mind. At the same time, major brands and retailers are adding environmental reporting requirements for their suppliers. Therefore, actors across the US dairy industry are working together to improve environmental performance for the entire supply chain and towards that end commissioned this study.

Life cycle assessment (LCA) is a technique for assessing the potential environmental impacts associated with a product, process, or service throughout its lifetime. LCAs have been used as a tool to identify "hot

Life Cycle Assessment of Cheese and Whey Production in the USA. © *Kim D, Thoma G, Nutter D, Milani F, Ulrich R, and Norris G.* The International Journal of Life Cycle Assessment *18,5 (2013). doi:10.1007/s11367-013-0553-9. Licensed under a Creative Commons Attribution License, http://creativecommons.org/licenses/by/3.0/.*

spots" in the production chain that may introduce opportunities for simultaneously lowering environmental impacts and improving efficiency and profitability (Eide 2002). This study is a cradle-to-grave LCA of natural cheese production focused on quantifying cumulative energy demand; emissions to air, water, and land; and consumption of water and other natural resources. There is a need to assess the impacts of these inventory flows on climate change, resource depletion, and human and ecosystem health.

Cheddar and mozzarella were chosen on the basis that they represent about 64% (by mass) and 80% (by sales) of all cheese produced in the USA (IDFA 2010). The principal objective of this work is to determine a baseline for the environmental impacts associated with production and consumption of cheddar and mozzarella cheese and associated whey products in the USA.

5.1.1 LITERATURE REVIEW AND BACKGROUND

Previous LCAs for dairy products have focused primarily on agricultural production (Cederberg and Mattsson 2000; Haas et al. 2001; Gerber et al. 2010). LCAs for the cheese industry are not extensive, but important research has been conducted in Australia (Lundie et al. 2003), Scandinavia (Berlin 2002; Dalgaard and Halberg 2004), and Western European countries (Bianconi et al. 1998; Hospido et al. 2003; Williams et al. 2006). In addition, considerable research has been done on food packaging, with some emphasis on milk packaging in particular (Keoleian and Spitzley 1999); however, no information regarding cheese packaging has been identified. Research on the life cycle of dairy products from retail to consumer to end-of-life has been minimal.

Many of the existing studies consider the footprint of milk leaving the farm; our review of the literature revealed few post-farm analyses. Sonesson and Berlin (2003) suggest that both packaging and transportation from the retail outlet to the home are major contributors. Other work by this group highlights the need for improvements in process management to minimize milk waste during processing of different products (Berlin 2005; Berlin et al. 2007). Nielsen and Høier (2009) have investigated yield improvement effects on environmental impacts of cheese production. In terms of overall global warming potential (GWP) of the supply chain, the majority of the effect originates from the farming activity, primarily from methane emissions from the cows and fertilizer production and use for feed. Another case study on Dutch cheese reached similar conclusions (van Middelaar et al. 2011). One recent study of US cheese production by Capper and Cady (2012) estimated the greenhouse gas (GHG) emissions from production of 500,000 tons of cheddar cheese derived from Jersey and Holstein milk, both with and without recombinant bovine somatotropin use. They report that cheddar cheese produced from Jersey milk had a lower footprint than cheese produced from Holsteins and use of recombinant bovine somatotropin further reduced GHG.

Several cradle-to-grave LCAs have been conducted for dairy products, including those previously noted. However, the scale, scope, and location were different from this study, and thus, they have limited direct applicability for assessing the US situation.

5.2 METHODS

This study has been structured following ISO 14040-compliant and ISO 14044-compliant LCA methodology (ISO 2006a, b). These standards provide an internationally agreed method of conducting LCA, but leave significant degrees of flexibility in methodology to customize individual projects.

5.2.1 GOAL AND SCOPE OF THE STUDY

The main goal of this work was to equip US cheese industry stakeholders with timely, defensible, and relevant information to support the incorporation of environmental performance into decision-making and support the development of innovative products, processes, and services. The study will provide cheese manufacturers an opportunity to benchmark their individual performance against a 2009 industry average, which is reported in this paper.

The scope of the project was a cradle-to-grave assessment with particular emphasis on the unit operations under direct control of a typical cheese-processing plant. In particular, these unit operations were transport of raw milk to the plant, cheese and whey manufacture, and delivery of cheese and whey products to the first customer.

5.2.2 FUNCTIONAL UNIT

Because cheese is produced with variable moisture content, the results are presented on a moisture-free basis. Three relevant functional units were defined:

- One ton (1,000 kg) of cheddar cheese consumed (dry weight basis);
- One ton of mozzarella cheese consumed (dry weight basis);
- One ton of dry whey delivered (dry weight basis).

5.2.3 SYSTEM BOUNDARIES AND CUTOFF CRITERIA

System boundaries encompass production of raw milk (feed production and on-farm), cheese manufacturing, packaging, transport, retail, consumption, and end-of-life (Fig. 1). We also analyzed the gate-to-grave system to increase resolution of the manufacturing and use phases. The boundary for whey does not include retail or consumption due to lack of data. We did not include in the inventory processes activities such as employee commuting; air travel; and veterinary, accounting, or legal services.

In determining whether to expend project resources to collect data for the inclusion of specific inputs, a 1% cutoff threshold for mass and energy was adopted. Although the study is intended to be comprehensive in consideration of impacts resulting from cheese supply chains, it is not a detailed engineering analysis of specific unit operations within the manufacturing sector. Thus, for example, we did not assign a specific energy requirement for cheese-making vats, cleaning in place, or starter culture operations, rather, we used the information available at the manufacturing plant scale, coupled with allocation of burdens to multiple plant products, to define the burden assigned to cheese, whey, and other coproducts. For this reason, it is important to state that all operations, as well as facility overhead (computers, heating, lights, etc.), are accounted for in this work.

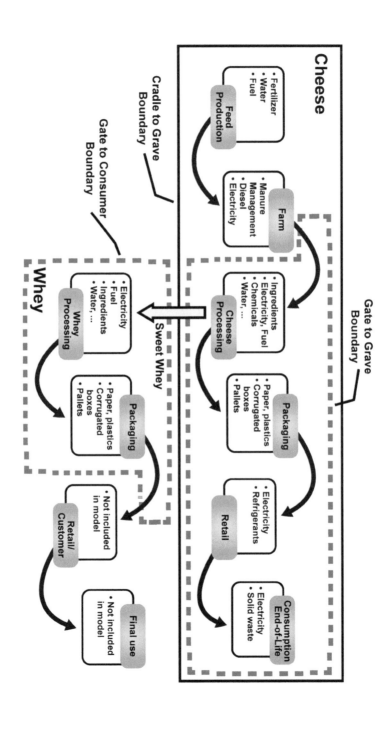

FIGURE 1: Flow diagram depicting cheese and whey unit processes/operations, applicable to both cheddar and mozzarella processes. Note that the curved arrows represent a transport operation

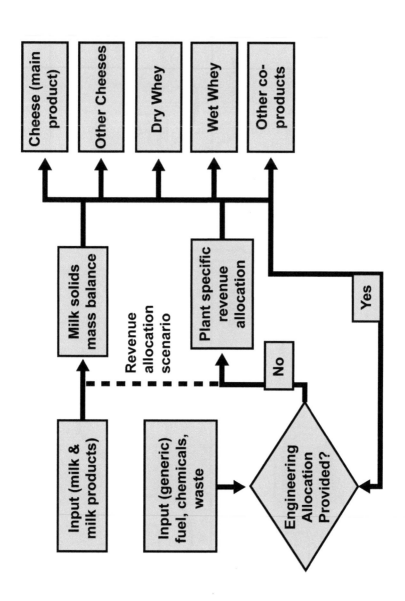

FIGURE 2: System separation for unique processes; milk solids, plant engineering estimates, and revenue-based approaches can be taken to allocate the common process burdens (Aguirre-Villegas et al. 2012)

5.2.4 ALLOCATION

Milk is the most significant input in the manufacturing of cheese, and milk solids (4.9% lactose, 3.4% fat, 3.3% protein, and 0.7% minerals) represent the important fraction of raw milk (87.7% water and 12.3% solids) in terms of cheese production. The production burden (at the dairy farm gate) for milk can be wholly assigned to the solids without differentiation (i.e., protein and fat assigned the same farm gate burden—water is considered only as a carrier), and the solids flow can be conceptually separated and treated as distinct inputs to the manufacturing system, allowing the solids content to be used as the mechanism for assigning the incoming milk burden to each coproduct (Feitz et al. 2007; Aguirre-Villegas et al. 2012). Allocation of the incoming milk solid burdens associated across the multiple coproducts based on milk solid distribution among the coproducts was our default approach (Fig. 2). In the plant survey, we requested each manufacturing facility operator to estimate the allocation of common utilities (electricity, natural gas, steam, etc.) to different operations within the plant boundary. Where this information was provided, we used it for plant-specific allocation of these inputs. For facilities that produce several types of cheese but have inputs without clearly identified fraction, the revenue associated with these sales were used to allocate the burdens among the cheeses.

5.2.5 LIFE CYCLE INVENTORY

This LCA is comprehensive and includes all inputs to the dairy industry, from crop farming to the final disposition of the packaging at the end of the supply chain. However, the primary focus of this study was on processes within the control of cheese-manufacturing plants. For each participating plant, processing companies were asked to complete a spreadsheet-based survey to facilitate incorporation of the data. During 2010, data from 2009 operations were collected from a total of 17 processing plants, including 10 cheddar manufacturing facilities (0.55 million tons of cumulative production) and 6 mozzarella manufacturing facilities (0.35 million tons of cumulative production). The industry average life cycle inventory (LCI) data are available in the Electronic supplementary material of this paper. Based on US production estimates of 1.45 million tons/year of cheddar and 1.47 million tons/year of mozzarella (IDFA 2010), the study has a sample representing 38 and 24% of production, respectively. A variety of plant sizes are represented, with production ranging from 0.014 to 0.14 million tons of cheese/year. The survey requested facility-level data regarding purchases (materials and energy), production (cheese and other products), and emissions (solid and liquid waste streams). Previous work conducted by the investigators for the production of fluid milk to the farm gate was used as background for milk production (Thoma et al. 2012a, b). Data collected from primary sources were checked for validity by ensuring consistency of units for reporting and conversion as well as material balances to insure that all incoming milk solids are accounted for in products leaving the manufacturing facility. The ecoinvent pedigree matrix approach to assigning uncertainty of inputs was applied to unit processes generated from primary data. Secondary data were taken from the ecoinvent v2.2 database. The data quality pedigree provided by the ecoinvent center for these data was adopted without revision. If secondary data are not available, input–output LCI datasets from the Open IO database were used as a proxy (TSC, Open IO) (TSC 2012). SimaPro© 7.3 (PRé Consultants, The Netherlands 2012) was used as the primary modeling software; the ecoinvent database, modified to account for US electricity, provided information on the "upstream" burdens associated with materials such as fuels and plant chemicals.

5.2.5.1 CHEESE AND WHEY PLANT DATA COLLECTION

The plants within the study combined cheese and whey production. The survey requested information at the subfacility scale; however, in many cases, only facility-level data were available. For example, most plants reported a single annual electrical energy use. We requested engineering estimates for separate material and/or energy flows (inputs and outputs) associated solely with cheese or whey products. This information was used in the algorithm that allocated material and energy flows between the coproducts of cheese and whey (see Fig. 2). For this study, each output product of each plant was classified as one of the following: main cheese, other cheese, dry whey, wet whey, and other coproducts. Protection of confidential business information requires an aggregation of the data that were acquired from the manufacturing facilities that participated in this project. Representative average production LCI data were generated using the allocated LCI data for each plant, which was totaled to create a generic inventory for each of the five potential coproducts at each facility. The resultant inventory is a production-weighted dataset because each product's reference flow was the sum of production from all reporting facilities.

5.2.5.2 TRANSPORTATION: FARM TO MANUFACTURING AND MANUFACTURING TO RETAIL

The survey included information on transportation distances from farm to manufacturing facility and also for distribution to retail (or in the case of whey, to the first customer). These data were used to determine the impacts of these stages within the cheese supply chain. The baseline vehicle was considered to be an insulated tanker truck and a refrigerated truck for raw milk and finished product, respectively. For a refrigerated truck transport, we modified ecoinvent unit process by adding refrigerant loss (Nutter et al. 2012). Empty kilometers (during return) were also included. Allocation of transportation of the raw milk to different products was based on milk solids. Post-manufacturing transport was directly assigned to the product being transported.

5.2.5.3 RETAIL

Contribution to environmental impacts from the retail sector was assessed from information previously requested from the project sponsor, who provided data regarding shelf space occupied in retail grocery outlets coupled with publicly available data for energy consumption in the building sector (Energy Information Administration (EIA) 2003; Energy Star 2008). Disposal of secondary packaging was accounted for through recycling rates of the materials commonly recycled (corrugated packaging and pallets). After distribution from the processor to the retail gate, cheese is displayed for consumer purchase. During this phase, there are three distinct emissions streams: refrigerant leakage, refrigeration electricity, and overhead electricity. For the purposes of this LCA, cheese sales channels were divided into two primary channels: supermarkets and mass merchandisers. Estimates of the sales volume, space occupancy, and energy demands were used to determine the burden of this supply chain stage.

5.2.5.4 CONSUMER TRANSPORTATION

According to the US Department of Transportation, Bureau of Transportation Statistics, Research and Innovative Technology Administration's 2001 National Household Travel Survey (NHTS), the average household makes 88.4 trips annually of 10.8 km roundtrip for shopping (NHTS 2009). Allocation of impacts from this activity to cheese is 1.16% (all cheese, 11.9% of dairy (USDA 2010); all dairy, 9.8% of grocery sales (Food Marketing Institute 2010)) resulting in 0.15 km/kg cheese. The average number of people per household is assumed to be 2.6 (US Census Bureau 2009), and the per capita annual consumption of cheese is estimated to be 3.69 and 3.95 kg for cheddar and mozzarella, respectively. Therefore, the total annual household consumption was estimated to be 9.59 and 10.26 kg for cheddar and mozzarella, respectively. Considering all cheeses (excluding ricotta and cottage cheeses), annual household consumption is calculated at 28.9 kg, resulting in 33.2 and 35.5% of all cheese consumption for cheddar and mozzarella, respectively. The transportation distances, then allocated to cheddar and mozzarella, are thus 0.050 and 0.054 km/kg cheese purchased, respectively.

The average fuel economy for passenger cars and other four-wheel vehicles (pickup truck, sport utility vehicles) was determined from the NHTS (2009) to be 9.61 and 7.70 km/L, respectively. It was assumed that all personal vehicles are powered by gasoline. The National Automobile Dealers Association (NADA) State of the Industry Report (NADA 2011) reports a 50:50 market share ratio of passenger cars to other four-wheel vehicles. Therefore, a weighted average of 8.65 km/L was assumed as the average fuel economy of personal vehicles. As the LCI datasets for personal transport in ecoinvent do not exactly match this fuel economy, we adjusted the number of kilometers of operation to ensure that the estimated fuel consumed, based on average US fuel economy, was properly calculated.

5.2.5.5 HOME REFRIGERATION

The EIA Residential Energy Consumption Survey estimates annual energy use for home refrigeration to be approximately 1,350 kWh (EIA 2005). The cheese portion of the total refrigerated products, 2.57% (cheese, 11.9% of all dairy (USDA 2010); dairy, 21.7% of refrigerated sales (Food Marketing Institute 2010)), is used to calculate the home refrigeration attributable to cheese, which results in 34.7 kWh. Note that the refrigerated shelf space allocation at home is expected to be an overestimate: in-home shelf space occupied by cheese is likely smaller than at the store, since the fraction of shelf space occupied in-home is likely decreased due to items purchased at the store unrefrigerated that need refrigeration upon opening (e.g., ketchup). With these caveats, refrigeration energy per kilogram of all cheese at household is then estimated to be 1.2 kWh/kg, and thus, 0.40 and 0.43 kWh/kg for cheddar and mozzarella, respectively, based on their market share.

5.2.5.6 DISHWASHING

Water and energy burdens for dishwashing were taken from the Energy Star criteria for a standard-sized dishwasher model (Energy Star 2009). A standard-sized model is considered to use 1.51 kWh and 22 L of water/cycle. It has a capacity of eight place settings and six serving pieces. A "place setting" is assumed to be comprised of two plates, one bowl, six utensils, and three glasses. Therefore, each cycle is assumed to wash 36 utensils (6 utensils × 6 serving pieces) and 48 non-utensils (6 non-utensils × 8 place settings). It is assumed that 10% of water and energy is allocated to the utensil rack and 90% to the non-utensil pieces. We assumed the dishwashing burden for utensils and plates for cheese consumption to be 5% of a dishwasher load/kg cheese consumed. This assumption was based on an estimate of the mean number of plates and utensils used for cheese and the capacity of a typical dishwasher.

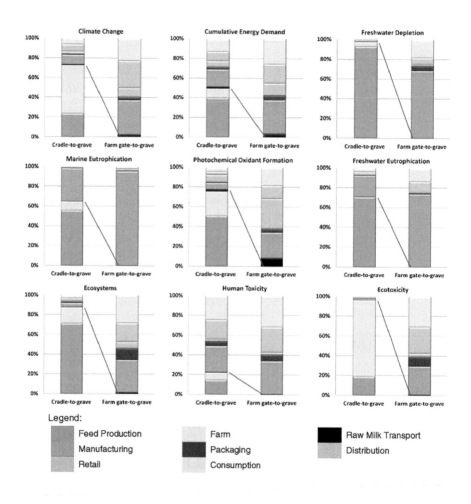

FIGURE 3: Contribution analysis of cradle-to-grave (left column) and farm gate-to-grave (right column) LCIA results for cheddar cheese supply chain

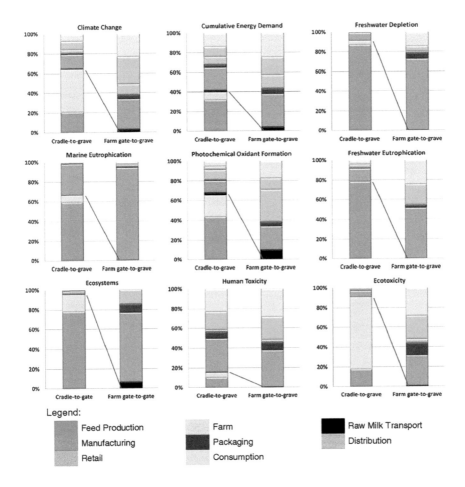

FIGURE 4: Contribution analysis of cradle-to-grave (left column) and farm gate-to-grave (right column) LCIA results for mozzarella cheese supply chain

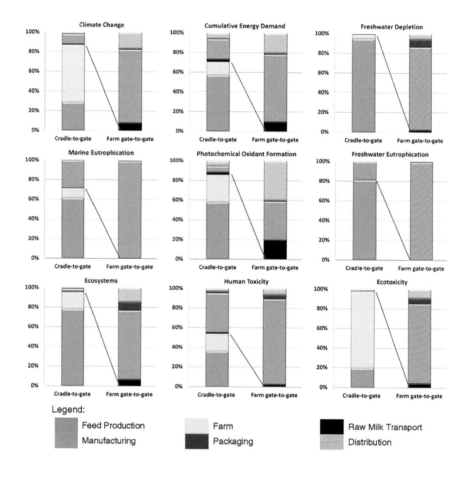

FIGURE 5: Contribution analysis of cradle-to-customer-gate (left column) and farm gate-to-customer-gate (right column) LCIA results for dry whey supply chain

Table 2: LCIA results per ton of cheddar cheese consumed (dry solids basis) from the cradle-to-grave

Impact category	Unit	Feed production	Farm	Raw milk transport	Manufacturing	Packaging	Distribution	Retail	Consumption	Total
Climate change	kg CO_2e	3.17E+03	6.77E+03	9.72E+01	1.09E+03	1.23E+02	3.38E+02	1.01E+03	8.39E+02	1.34E+04
Cumulative energy demand	MJ	2.97E+04	7.59E+03	1.39E+03	1.37E+04	1.94E+03	4.83E+03	7.48E+03	1.01E+04	7.67E+04
Freshwater depletion	m³	1.27E+03	6.00E+01	8.23E-02	2.68E+01	1.36E+00	2.86E-01	1.75E+00	4.73E+00	1.37E+03
Marine eutrophication	kg N eq.	2.13E+01	3.48E+00	3.60E-02	1.26E+01	9.10E-02	1.25E-01	1.15E-01	2.48E-01	3.80E+01
Photochemical oxidant formation	kg NMVOC	2.25E+01	1.22E+01	1.02E+00	3.26E+00	4.62E-01	3.55E+00	1.50E+00	2.16E+00	4.67E+01
Freshwater eutrophication	kg P eq.	6.34E+00	1.40E-02	9.63E-04	7.86E-01	2.93E-02	3.35E-03	3.03E-01	3.55E-01	7.84E+00
Ecosystems	Species/year	2.54E-04	5.75E-05	7.76E-07	1.41E-05	5.08E-06	2.70E-06	8.18E-06	1.20E-05	3.55E-04
Human toxicity	CTUh	6.82E-05	3.80E-05	2.42E-06	1.36E-04	2.62E-05	8.43E-06	1.02E-04	1.25E-04	5.07E-04
Ecotoxicity	CTUe	1.35E+04	5.57E+04	1.98E+01	1.08E+03	2.03E+02	6.88E+01	5.56E+02	6.32E+02	7.17E+04

TABLE 3: LCIA results per ton of mozzarella cheese consumed (dry solids basis) from the cradle-to-grave

Impact category	Unit	Feed production	Farm	Raw milk transport	Manufacturing	Packaging	Distribution	Retail	Consumption	Total
Climate change	kg CO_2e	2.90E+03	6.20E+03	1.50E+02	2.02E+03	2.09E+02	4.73E+02	1.24E+03	1.03E+03	1.42E+04
Cumulative energy demand	MJ	2.72E+04	6.95E+03	2.14E+03	2.05E+04	3.04E+03	6.77E+03	9.19E+03	1.24E+04	8.83E+04
Freshwater depletion	m^3	1.17E+03	5.49E+01	1.27E-01	1.01E+02	2.63E+00	4.01E-01	2.13E+00	5.81E+00	1.33E+03
Marine eutrophication	kg N eq.	1.95E+01	3.19E+00	5.54E-02	1.44E+01	1.73E-01	1.75E-01	1.41E-01	3.04E-01	3.80E+01
Photochemical oxidant formation	kg NMVOC	2.06E+01	1.12E+01	1.57E+00	6.08E+00	7.54E-01	4.97E+00	1.85E+00	2.66E+00	4.97E+01
Freshwater eutrophication	kg P eq.	5.81E+00	1.28E-02	1.48E-03	1.36E+00	5.33E-02	4.69E-03	3.72E-01	4.37E-01	8.05E+00
Ecosystems	Species/year	2.33E-04	5.27E-05	1.19E-06	3.43E-05	9.23E-06	3.78E-06	9.99E-06	1.47E-05	3.59E-04
Human toxicity	CTUh	6.25E-05	3.48E-05	3.73E-06	2.27E-04	4.60E-05	1.18E-05	1.25E-04	1.54E-04	6.65E-04
Ecotoxicity	CTUe	1.24E+04	5.10E+04	3.05E+01	4.63E+03	3.54E+02	9.64E+01	6.54E+02	7.77E+02	6.99E+04

5.2.5.7 POSTCONSUMER SOLID WASTE

We model waste disposal in SimaPro© with unit processes from ecoinvent for consumer disposal of packaging material. Franklin Associates (2008) report that an estimated 14% of postconsumer waste is incinerated with energy recovery. We modeled the incineration of these materials but did not account for energy recovery, as it fell below the 1% cutoff criterion.

TABLE 1: Reporting categories used for the study

Life cycle inventory categories	Life cycle impact categories
Cumulative energy demand	Climate change
Freshwater depletion	Marine eutrophication
	Photochemical oxidant formation
	Freshwater eutrophication
	Ecosystems
	Human toxicity
	Ecotoxicity

5.2.6 SCENARIO ANALYSIS OF CHEDDAR AGING

The bulk of cheddar cheese sold in the USA is aged approximately 70 days, but specialty cheddar can be aged five or more years. In 2009, 1.45 million tons of cheddar cheese was produced in the USA (IDFA 2010). Cold holding reports for cheddar cheese were examined and a typical inventory of 0.28 million tons was reported (NASS 2010). Using a simple first in–first out assumption, the US inventory of cheddar cheese turns over 5.17 times a year (1.45/0.28=5.17), implying that the typical age of cheddar cheese at retail is 70.6 days (365 days/5.17=70.6 days). Based on EIA (2003) survey data, refrigerated warehouses consume an average of 307 kWh/m2/year of electricity and 338 MJ/m^2/year of natural gas. We assumed that pallets were stored on shelves up to six pallets high (typical warehouse height=~9 m) and used an industry estimate of the number of 18.1 kg (40-lb) blocks in 45 blocks per pallet. Ammonia is used for refrigeration in large warehouses used for cheese storage. We used an emission factor of 13.6 kg NH3/employee/year coupled with Industrial Assessment Center (IAC 2009) data on employees and warehouse size to arrive at an estimated emission of 0.013 kg NH_3/m^2/day of storage. Mozzarella is distributed for retail as rapidly as possible, but typically, needs to be held for 2 weeks before unwrapping to smaller pieces and repackaging to retail sizes. It should be noted that a large fraction of mozzarella is used in food service applications where it is frozen and stored for some time prior to being used. We did not include this branch of the supply chain as our focus was on cheese directly purchased by the end consumer.

5.2.7 LIFE CYCLE IMPACT ASSESSMENT

The intention of this study was to provide a comprehensive environmental life cycle impact assessment (LCIA) of cheese production and consumption, which stems from all phases of cheese production and delivery systems. These environmental impacts include climate change, cumulative energy demand, freshwater depletion, marine and freshwater eutrophication, photochemical oxidant formation, impacts to ecosystems and human toxicity, and ecotoxicity (Hertwich et al. 1998; Huijbregts et al. 2000; Jolliet et al. 2003; Goedkoop et al. 2009; Hischier and Weidema 2010). We chose impact categories relevant to the dairy industry: IPCC GWP 100a, Cumulative Energy Demand, ReCiPe Midpoint, ReCiPe Endpoint, and USEtox (Table 1).

5.3 RESULTS AND DISCUSSION

5.3.1 LIFE CYCLE IMPACT ASSESSMENT RESULTS

We accounted for the entire supply chain of cheese consumption in the USA. This includes specifically product loss at various stages of the supply chain, as well as consumer transport and storage of products prior to consumption. Figures 3 and 4 present a contribution analysis of cradle-to-grave (left column) and farm gate-to-grave (right column) LCIA results across all of the impact categories considered for cheddar and mozzarella, respectively. Figure 5 presents a summary of cradle-to-gate and farm gate-to-gate LCIA results for dry whey. Quantitative results are presented in Tables 2 and 3 for cheddar and mozzarella, respectively. It is not surprising that, for most of the impact categories, the production of milk dominates the environmental impacts of cheese production. Cumulative energy demand and human toxicity are the only two categories for which 50% or more of the impact occurs after the farm gate. This can be explained by the relatively even distribution of electricity consumption across the supply chain compared to the more intense pre-farm gate activities that affect the remaining impact categories.

TABLE 4: The largest impact drivers for feed production and on-farm activities

Impact category	Impact drivers
Climate change	Farm-based enteric methane and manure management; farm CO_2 from fossil fuels combustion (cultivation and on-farm usage); and N_2O from fertilizer application and manure management
Cumulative energy demand	Natural gas, oil, and coal for direct use and production of nitrogen fertilizer
Freshwater depletion	Majority of irrigation (95%) and lesser amount toward milking parlor cleaning and livestock watering
Marine eutrophication	Phosphate release from runoff due to on-field fertilizer application
Photochemical oxidant formation	NOx and VOCs from combustion (a significant source in some regions is ethanol released during fermentation of silage)
Freshwater eutrophication	Nitrogen compound runoff from fertilizer application and manure management; eutrophication is geospatially variable and dependent on local conditions
Ecosystems	Land occupation for crop production and crop/farm GHG emissions; land occupation is often considered a surrogate indicator for biodiversity
Human toxicity	Arsenic to water and heavy metals (in both air and water) primarily from coal mining tailings and coal ash disposal in the electricity supply chain
Ecotoxicity	Pesticides for crop and livestock protection; insecticide applied as a back pour for fly and lice control in dairy cattle contributes significantly of the total aquatic ecotoxic impact

5.3.1.1 FEED PRODUCTION AND ON-FARM IMPACTS

We found that raw milk impacts from feed production and farm milk production are similar for each manufactured product (cheddar, mozzarella, and dry whey). Both are significant contributors across all impact categories (Table 4).

5.3.1.2 FARM GATE-TO-GATE IMPACTS

Environmental impacts associated with foreground LCA processes are those that can be more readily controlled by the cheese manufacturer and span the supply chain from the dairy farm gate to the first customer. It is noted that impacts in climate change and cumulative energy demand are significantly driven by electricity and natural gas consumption (Fig. 6). Cumulative energy demand normally tracks GHG emissions. However, milk production has a significant contribution from enteric methane, and therefore, the relative consumption of fossil fuel is lower. Eutrophication impacts are dominated by on-site wastewater treatment (WWT). In this model, WWT is taken from the ecoinvent database and assumes a loading of nitrogen and phosphorus that may not be representative of individual plants in the US industry. Some facilities reported significantly lower phosphorous loadings, while others reported significantly higher loadings. Based on the survey data and literature reports, we modified the ecoinvent dataset to remove contributions from sludge incineration, to the extent that it could be singled out, and reduced the estimated influent phosphorus concentration from 250 mg/L (Swiss conditions) to 70 mg/L. It is necessary to have more site-specific information to draw conclusions about individual facilities. Photochemical oxidant formation is strongly influenced by transportation. The impacts to human toxicity and ecotoxicity are dominated by electricity use (arsenic and other heavy metals emissions from coal mining activities). The relative contributions to cheddar and mozzarella impacts are nearly identical on a dry solids basis because the technologies are fairly similar. Thus, mozzarella-specific results are not included here.

5.3.2 NORMALIZATION

This is an optional phase of an LCA according to ISO 14044, but it is a useful step to help identify impact categories that are particularly relevant for

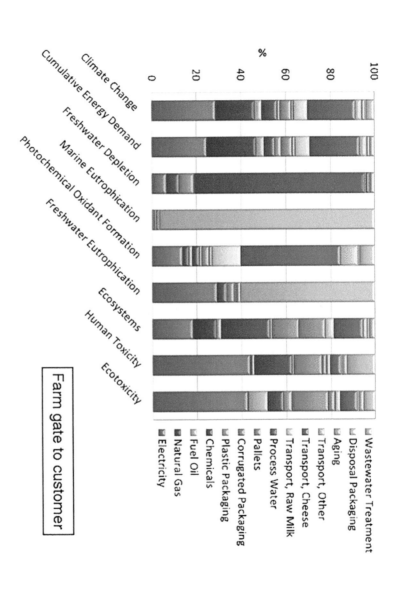

FIGURE 6: Relative contribution of manufacturing inputs to environmental impacts from farm gate-to-customer for cheddar cheese production

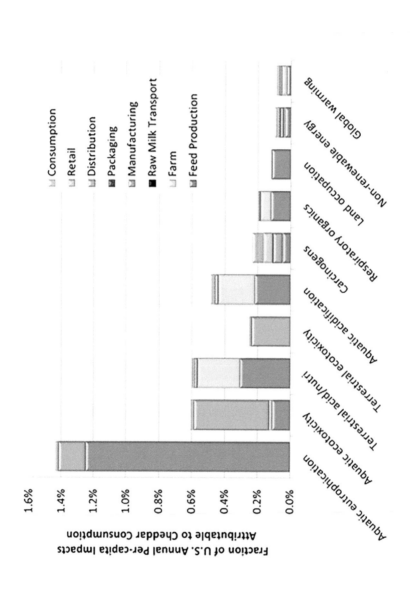

FIGURE 7: US normalization of cradle-to-grave impacts for the consumption of 3.7 kg (63.2% solids) cheddar cheese using normalization factors provided by Lautier et al. (2010) for the IMPACT 2002+ LCIA framework

the industry under study. Briefly, normalization is an effort to contextualize the emissions impacts, typically on a regional basis. The total emissions contributing to a specific impact category for the region are estimated and then normalized to a per person basis for the region. Thus, a normalization factor will have units of impact per person per year and represents the average cumulative impacts in the region on a per capita basis. Normalization factors have been recently published for the USA for IMPACT 2002+ (Lautier et al. 2010). Comparison of impact categories from different methods is a valuable qualitative exercise. However, caution should be taken in making direct quantitative comparisons because there are differences in the underlying methods, and characterization of the same substance in different frameworks is not exactly the same. Some of these differences arise due to the geographic domain of the framework, while other differences represent uncertainty in, for example, the toxicology of a particular compound or whether a receiving water body is phosphorus-limited or nitrogen-limited in the case of eutrophication. In the following discussion, each of two normalization tests was conducted using the reported (USDA 2010) annual loss-adjusted cheese consumption of an average US citizen, 3.67 kg cheddar (2.32 kg on a dry solids basis) or 3.95 kg mozzarella (2.03 kg on a dry solids basis).

5.3.2.1 IMPACT 2002+ US MIDPOINT NORMALIZATION

We conducted a normalization test using the IMPACT 2002+ assessment framework. The emissions from the system are compared to the average per capita emissions, enabling mitigation efforts to focus on the impact categories that contribute the largest relative fraction of environmental impact. There is not an exact correspondence between the IMPACT 2002+ framework and the combination of ReCiPe and USEtox chosen for this study, but US normalization factors do not exist for those methods. The example normalization results for cheddar (Fig. 7) indicate that aquatic eutrophication, aquatic ecotoxicity, and terrestrial acidification are important categories on which to focus improvement activities. From a manufacturing perspective, these can be mitigated through energy conservation and water conservation/treatment activities.

5.3.2.2 RECIPE WORLD ENDPOINT NORMALIZATION

The ReCiPe method presents an alternate view of normalization of cheddar impacts (Fig. 8). In this approach, the region under consideration is the globe. Thus, the y-axis represents the impact that a US resident's consumption of cheese imposes on the environment compared to an average impact of all people on the planet (from all sources contributing to that impact category). It is well-documented that the USA consumes a disproportionate amount of resources compared to the majority of the world. The global normalization approach, based on an endpoint perspective, which accounts for the effects of, for example, climate change on human health, suggests that US annual consumption of natural cheese is a more important driver of climate change health effects, respiratory effects resulting from particulate-forming emissions and fossil fuel depletion, with relatively lower importance in the remaining categories. It should be noted that the set of impacts two orders of magnitude smaller than the others (see Fig. 8) should have lower priority for reduction. It is of course important to make incremental improvements in all impact categories, and efforts to reduce electricity and fossil fuel consumption will have broad benefits.

5.3.3 SCENARIO ANALYSIS OF CHEDDAR AGING

To understand the potential impacts associated with long-term aging of cheddar, we conducted a scenario study with cold storage up to 5 years. We present the results for both cradle-to-grave and post-farm gate (Table 5). In terms of GHG emissions after 60 months of aging, there is approximately 6 and 22% increase (an increase of 0.47 kg CO_2e/kg dry cheese solids) in the cradle-to-grave and farm gate-to-grave emissions, respectively. For the post-farm supply chain, human toxicity and ecotoxicity impacts increase noticeably, primarily associated with increased electricity use for additional refrigeration.

TABLE 5: Percent increase in cradle-to-grave and post-farm gate impacts (in parentheses) from 1 to 60 months cheddar aging in refrigerated warehouse (the baseline scenario assumed ~70.6 days total aging, if manufacturing plants hold 10–14 days—values in this table are for total aging time)

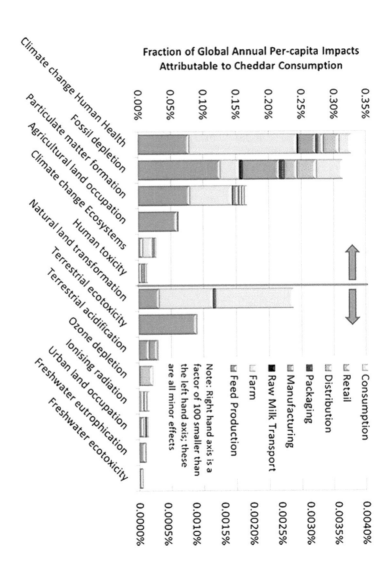

FIGURE 8: Normalized cradle-to-grave impacts for the consumption of 3.7 kg cheddar by US consumers. ReCiPe Endpoint impacts with World Hierarchist normalization

Impact category	Months of additional aging,% (%)						
	1	2	12	24	36	48	60
Climate change	0.09 (0.37)	0.19 (0.73)	1.1 (4.4)	2.2 (8.8)	3.3 (13)	4.5 (18)	5.6 (22)
Cumulative energy demand	0.22 (0.43)	0.44 (0.87)	2.6 (5.2)	5.3 (10)	7.9 (16)	11 (21)	13 (26)
Freshwater depletion	0.00 (0.13)	0.00 (0.25)	0.03 (1.5)	0.06 (3.0)	0.08 (4.5)	0.11 (6.0)	0.14 (7.5)
Marine eutrophication	0.06 (0.18)	0.12 (0.36)	0.75 (2.2)	1.5 (4.4)	2.2 (6.6)	3.0 (8.7)	3.7 (11)
Photochemical oxidant formation	0.06 (0.26)	0.13 (0.52)	0.77 (3.1)	1.5 (6.2)	2.3 (9.3)	3.1 (12)	3.9 (15)
Freshwater eutrophication	0.07 (0.41)	0.15 (0.83)	0.89 (5.0)	1.8 (9.9)	2.7 (15)	3.6 (20)	4.5 (25)
Ecosystems	0.03 (0.26)	0.06 (0.52)	0.36 (3.1)	0.71 (6.2)	1.1 (9.4)	1.4 (12)	1.8 (16)
Human toxicity	0.39 (0.50)	0.78 (1.0)	4.7 (6.0)	9.4 (12)	14 (18)	19 (24)	23 (30)
Ecotoxicity	0.01 (0.50)	0.03 (1.0)	0.17 (6.0)	0.34 (12)	0.51 (18)	0.69 (24)	0.86 (30)

5.3.4 UNCERTAINTY ANALYSIS

LCA results for 1 ton of cheddar, mozzarella, and dry whey consumption in 2009 were analyzed using 1,000 Monte Carlo analysis runs each (Tables 6, 7, and 8, respectively). GHG emissions are of notable interest, and on dry milk solids basis, the carbon footprint of cheddar and mozzarella are approximately 13.4 and 14.2 tons CO_2e/ton of cheese solids consumed, respectively. The 95% confidence interval (CI) ranges 9.28–19.3 tons CO_2e/ton of cheddar solids consumed and 9.73–19.6 tons of CO_2e/ton of mozzarella cheese solids consumed. For an average moisture content of 36.8% for cheddar as sold at retail, the carbon footprint is 8.60 tons CO_2e/ton of cheddar cheese consumed (95% CI = 5.86–12.2). Based on an average moisture content of 48.6% for mozzarella as sold at retail, the carbon footprint is 7.28 tons CO_2e/ton of mozzarella consumed (95% CI = 5.13–9.89). Freshwater depletion—defined as water removed during production but not returned to the same watershed, which excludes process and cooling water—is dominated by feed production due to crop irrigation. On a milk solids basis, freshwater depletion is 1,370 m³/ton of cheddar consumed (95% CI = 890–2,060 m³). This is equivalent to approximately 870 L of water/kg of cheddar cheese (as sold at retail) consumed in the USA. For mozzarella cheese, 680 L of water are consumed/kg of mozzarella consumed because it has higher moisture content at retail.

TABLE 6: Results of 1,000 Monte Carlo simulations for uncertainty analysis of cheddar cheese from cradle-to-grave per ton of dry cheese solids

Impact category	Unit	Mean	Coefficient of variation (CV) (%)	95% CI	
Climate change	kg CO₂e	1.34E+04	18.8	9.28E+03	1.93E+04
Cumulative energy demand	MJ	7.67E+04	46.1	4.73E+04	1.57E+05
Freshwater depletion	m³	1.37E+03	21.8	8.88E+02	2.06E+03
Marine eutrophication	kg N eq.	3.80E+01	18.3	2.65E+01	5.42E+01
Photochemical oxidant formation	kg NMVOC	4.67E+01	17.7	3.22E+01	6.54E+01
Freshwater eutrophication	kg P eq.	7.84E+00	18.7	2.41E−04	5.15E−04
Ecosystems	Species/year	3.55E−04	27.7	4.95E+00	1.22E+01
Human toxicity	CTUh	5.07E−04	211	1.81E−04	1.38E−03
Ecotoxicity	CTUe	7.17E+04	22.4	4.78E+04	1.06E+05

5.3.5 PLANT-SCALE VARIABILITY

Figure 9 presents a summary comparison of the ten cheddar and six mozzarella plants that provided data for this study to present the full variability of operations. The LCIA was normalized so that the average mozzarella plant equaled 100. Caution in interpreting this variability is necessary because each facility is considered in its totality, with no allocation between multiple products. The impacts are based on a reference flow of the total milk solids processed regardless of the quantity of cheese, whey, or other products manufactured. It is apparent that there is significant variability among the plants and that there are opportunities for many of them to improve. Due to the nature of the survey data collected (i.e., at the plant scale), it is not possible in this study to identify which unit operations may be causing the differences among the facilities, and additional subfacility data collection and analysis will be necessary to fully identify and target unit operations for improvement.

TABLE 7: Results of 1,000 Monte Carlo simulations for uncertainty analysis of mozzarella cheese from cradle-to-grave per ton of dry cheese solids

Impact category	Unit	Mean	CV (%)	95% CI	
Climate change	kg CO_2e	1.42E+04	17.0	9.73E+03	1.96E+04
Cumulative energy demand	MJ	8.83E+04	41.1	5.01E+04	1.69E+05
Freshwater depletion	m^3	1.33E+03	20.1	8.81E+02	1.94E+03
Marine eutrophication	kg N eq.	3.80E+01	16.9	2.64E+01	5.20E+01
Photochemical oxidant formation	kg NMVOC	4.97E+01	16.7	3.45E+01	6.65E+01
Freshwater eutrophication	kg P eq.	8.05E+00	16.8	2.47E−04	4.84E−04
Ecosystems	Species/year	3.59E−04	26.7	4.99E+00	1.26E+01
Human toxicity	CTUh	6.65E−04	95.2	2.57E−04	1.71E−03
Ecotoxicity	CTUe	6.99E+04	20.1	4.60E+04	9.98E+04

5.3.6 LIMITATIONS

The variability associated with the allocation procedure used for the study places some limits on the recommendations that can be supported for specific products. This work is intended to provide a benchmark for the industry, and for whole-plant analysis, this has been achieved. However, there were variations in the reported allocation of in-plant resource use. Without a detailed process model for each plant, a generic allocation is difficult to achieve. In addition, some data quality concerns exist regarding the completeness of the milk solid mass fractions. Thus, while the results are in good general agreement with available European studies, public statements regarding the footprint of cheese alone or whey alone should be made cautiously due to the individual allocation fractions of each plant studied.

TABLE 8: Results of 1,000 Monte Carlo runs for uncertainty analysis of dry whey from cradle-to-customer per ton of dry whey solids

Impact category	Unit	Mean	CV (%)	95% CI	
Climate change	kg CO_2e	1.21E+04	15.3	9.11E+03	1.61E+04
Cumulative energy demand	MJ	5.81E+04	28.5	4.09E+04	8.93E+04
Freshwater depletion	m^3	1.45E+03	16.2	1.05E+03	2.00E+03
Marine eutrophication	kg N eq.	3.73E+01	12.2	2.92E+01	4.77E+01
Photochemical oxidant formation	kg NMVOC	4.40E+01	12.9	3.33E+01	5.60E+01
Freshwater eutrophication	kg P eq.	7.52E+00	15.6	5.53E+00	1.01E+01
Ecosystems	Species/year	3.51E−04	13.4	2.70E−04	4.54E−04
Human toxicity	CTUh	2.27E−04	116	7.78E−05	7.29E−04
Ecotoxicity	CTUe	7.57E+04	14.9	5.69E+04	1.01E+05

5.4 CONCLUSIONS AND RECOMMENDATIONS

This study was a US-based LCA for cheddar cheese, mozzarella cheese, and associated whey products. A combination of LCI and LCIA were reported to help the cheese industry engage in sustainable practices and reduce environmental impacts, while validating those reductions by benchmarking their performance against a 2009 industry average. Primary focus was placed on the processes within the control of cheese-manufacturing plants.

Climate change and cumulative energy demand impacts are closely linked to fossil fuel consumption. Moreover, many other environmental impacts from the post-farm manufacturing and distribution stages of the production of cheddar cheese and mozzarella cheese are also directly linked to energy consumption primarily that associated with coal mining and combustion.

The production of raw milk is the major contributor to nearly all impact categories; thus, efforts to reduce milk/cheese loss at all stages in the supply chain have significant potential to reduce the overall impacts of cheese consumption. In addition, on-farm mitigation efforts around enteric methane, manure management, phosphate and nitrate runoff, and pesticides used on crops and livestock also have the potential to significantly reduce overall impacts.

Water-related impacts such as depletion and eutrophication can be considered as resource management issues—in the case of depletion, management of water quantity, and in the case of eutrophication, management of nutrients. Thus, opportunities for water conservation across the supply chain should be evaluated, and cheese manufacturers, while not having control over the largest fraction of water consumption, can begin to investigate the water use efficiency of the milk they procure.

The regionalized normalization analysis based on an average US citizen's annual cheese consumption showed that eutrophication represents the largest relative impact due largely to the combination of phosphorus runoff from agricultural fields and phosphorous emissions associated with digestion of wastewater from whey processing. Therefore, incorporating best practices around phosphorous and nitrogen management could yield improvements.

Finally, the US electricity supply chain (primarily coal-based) and combustion of other fossil fuels (natural gas, diesel, etc.) were also found to be the primary contributors to photochemical oxidant formation, impacts to ecosystems, human toxicity, and ecotoxicity. Thus, conservation efforts to reduce fuel and electricity use within the cheese life cycle will have broad beneficial impacts, both economic, through cost savings, and environmental, due to the reduction in emissions.

REFERENCES

1. Aguirre-Villegas HA, Milani FX, Kraatz S, Reinemann DJ (2012) Life cycle impact assessment and allocation methods development for cheese and whey processing. A Soc Agricul Biol Eng 55(2):613–627
2. Berlin J (2002) Environmental life cycle assessment (LCA) of Swedish semi-hard cheese. Int Dairy J 12(11):939–953
3. Berlin J (2005) Environmental improvements of the post-farm dairy chain: production management by systems analysis methods. Dissertation, Chalmers University of Technology, Göteborg, Sweden

4. Berlin J, Sonesson U, Tillman AM (2007) A life cycle based method to minimise environmental impact of dairy production through product sequencing. J Cleaner Prod 15:347–356
5. Bianconi P, Marani S, Masoni P, Raggi A, Sara B, Scartozzi D, Tarantini M (1998) Application of life-cycle assessment to the italian dairy industry: a case-study. Proceedings of the International Conference on Life Cycle Assessment in Agriculture, Agro-Industry and Forestry, Bruxelles, Belgio, December, pp 59–62
6. Capper JL, Cady RA (2012) A comparison of the environmental impact of Jersey compared with Holstein milk for cheese production. J Dairy Sci 95(1):165–176
7. Cederberg C, Mattsson B (2000) Life cycle assessment of milk production—a comparison of conventional and organic farming. J Cleaner Prod 8(1):49–60
8. Dalgaard R, Halberg N (2004) LCA of Danish milk—system expansion in practice. DIAS Report, An Hus 61:285–288
9. EarthShift (2012) US-EI Database. Available at http://www.earthshift.com/software/simapro/USEI-database. Accessed 28 September 2012
10. Energy Information Administration (EIA) (2003) Commercial buildings energy consumption survey. US Energy Information Administration. Available at http://www.eia.gov/emeu/cbecs/cbecs2003/detailed_tables_2003/detailed_tables_2003.html#enduse03
11. Energy Information Administration (EIA) (2005) Residential Energy Consumption Survey (RECS), Consumption & expenditures, home appliances and lighting. Available at http://www.eia.gov/consumption/residential/data/2005/index.cfm#tabs-2. Accessed 19 June 2011
12. Eide HM (2002) Life cycle assessment (LCA) of industrial milk production. Int J Life Cycle Assess 7(2):115–126
13. Energy Star (2008) Building upgrade manual chapter 11: grocery and convenience stores. Available at http://www.energystar.gov/ia/business/EPA_BUM_Full.pdf
14. Energy Star (2009) Dishwashers key product criteria. Available at http://www.energystar.gov/index.cfm?c=dishwash.pr_crit_dishwashers
15. Feitz AJ, Lundie S, Dennien G, Morain M, Jones M (2007) Generation of an industry-specific physico-chemical allocation matrix. Application in the dairy industry and implications for systems analysis. Int J Life Cycle Assess 12(2):109–117
16. Food Marketing Institute (2010) Supermarket facts. Available at http://www.fmi.org/facts_figs/?fuseaction=superfact. Accessed 20 June 2011
17. Franklin Associates (2008) LCI summary for four half-gallon milk containers. Peer Reviewed Final Report, September 2008
18. Frischknecht R, Rebitzer G (2005) The ecoinvent database system: a comprehensive web-based LCA database. J Cleaner Prod 13(13–14):1337–1343
19. Gerber P, Vellinga T, Opoio C, Henderson B, Steinfield H (2010) Greenhouse gas emissions from the dairy sector: a life cycle assessment. A report of the Food and Agriculture Organization, 96 pp. Available at http://www.fao.org/docrep/012/k7930e/k7930e00.pdf. Accessed 13 September 2012
20. Goedkoop M, Heijungs R, Huijbregts M, Schryver AD, Struijs J, Jelm R (2009) ReCiPe 2008. A life cycle impact assessment method which comprises harmonized category indicators at the midpoint and the endpoint level. Available at http://www.lcia-recipe.net

21. Haas G, Wetterich F, Kopke U (2001) Comparing intensive, extensified and organic grassland farming in southern Germany by process life cycle assessment. Agri Ecosys Environ 83:43–53

22. Hertwich EG, Pease WS, McKone TE (1998) Evaluating toxic impact assessment methods: what works best? Environ Sci Tech 32:A138–A144

23. Hischier R, Weidema B (2010) Implementation of life cycle impact assessment methods, data v2.2. Ecoinvent Center, Swiss Centre for Life Cycle Inventories. Available at http://www.ecoinvent.org/fileadmin/documents/en/03_LCIA-Implementation-v2.2.pdf

24. Hospido A, Moreira MT, Feijoo G (2003) Simplified life cycle assessment of Galician milk production. Int Dairy J 13(10):783–796

25. Huijbregts MAJ, Thissen U, Guinee JB, Jager T, Van de Meent D, Ragas AMJ, Wegener SA, Reijnders L (2000) Priority assessment of toxic substances in life cycle assessment, I: calculation of toxicity potential for 181 substances with the nested multi-media fate, exposure and effects model USES-LCA. Chemosphere 41:541–573

26. IAC (2009) Industrial Assessment Centers Database. US Department of Energy, Washington, DC

27. IDFA (2010) Dairy facts. International Dairy Foods Association, Washington, DC

28. ISO (2006a) ISO 14040: environmental management—life cycle assessment—principles and framework. International Organization of Standardization, Geneva

29. ISO (2006b) ISO 14044: environmental management—life cycle assessment—requirements and guidelines. International Organization of Standardization, Geneva

30. Jolliet O, Margni M, Charles R, Humbert S, Payet J, Rebitzer G, Rosenbaum R (2003) IMPACT 2002+: a new life cycle impact assessment methodology. Int J Life Cycle Assess 8(6):324–330

31. Keoleian GA, Spitzley DV (1999) Guidance for improving life-cycle design and management of milk packaging. J Ind Ecol 3(1):111–126

32. Lautier A, Rosenbaum RK, Margni M, Bare J, Roy PO, Beschenes L (2010) Development of normalization factors for Canada and the United States and comparison with European factors. Sci Tot Env 409(1):33–42

33. Lundie S, Feitz A, Jones M, Dennien G, Morian M (2003) Evaluation of the environmental performance of the Australian dairy processing industry using life cycle assessment. Dairy Research and Development Corporation, Canberra

34. NASS (2010) National Agricultural Statistics Service, Wisconsin Agricultural Statistics, United States Department of Agriculture (USDA-NASS) in cooperation with Wisconsin Department of Agriculture, Trade and Consumer Protection (WDATCP), Washington, DC. Available at http://www.nass.usda.gov. Accessed 11 December 2011

35. NADA (2011) National Automobile Dealers Association Data, State of the Industry Report, NADA's Industry Analysis Division. Available at http://www.nada.org/nadadata

36. NHTS (2009) Federal Highway Administration. National Household Travel Survey. Available at http://nhts.ornl.gov

37. Nielsen P, Høier E (2009) Environmental assessment of yield improvements obtained by the use of the enzyme phospholipase in mozzarella cheese production. Int J Life Cycle Assess 14(2):137–143

38. Nutter DW, Kim D, Ulrich R, Thoma G (2012) Greenhouse gas emission analysis for USA fluid milk processing plants: processing, packaging, and distribution. Int Dairy J. doi:10.1016/j.idairyj.2012.09.011

39. SimaPro© 7.3, PRé Consultants, The Netherlands (2012) Available at http://www.pre.nl/

40. Sonesson U, Berlin J (2003) Environmental impact of future milk supply chains in Sweden: a scenario study. J Clean Prod 11:253–266

41. Thoma G, Popp J, Shonnard D, Nutter D, Matlock M, Ulrich R, Kellogg W, Kim DS, Neiderman Z, Kemper N, Adom F, East C (2012a) Regional analysis of greenhouse gas emissions from US dairy farms: a cradle to farm-gate assessment of the American dairy industry circa 2008. Int Dairy J. doi:10.1016/j.idairyj.2012.09.010

42. Thoma G, Popp J, Nutter D, Shonnard D, Ulrich R, Matlock M et al (2012b) Greenhouse gas emissions from milk production and consumption in the United States: a cradle to grave life cycle assessment circa 2008. Int Dairy J. doi:10.1016/j.idairyj.2012.08.013

43. TSC (2012) Open IO. The Sustainability Consortium. Available at http://www.sustainabilityconsortium.org/open-io/. Accessed 4 December 2012

44. US Census Bureau (2009) Available at http://www.census.gov/acs/www/data_documentation/2009_release/

45. USDA (2010) Economic Research Service. Loss-adjusted food availability. United States Department of Agriculture. Available at http://www.ers.usda.gov/Data/Food-Consumption/FoodGuideSpreadsheets.htm

46. van Middelaar CE, Berentsen PBM, Dolman MA, de Boer IJM (2011) Eco-efficiency in the production chain of Dutch semi-hard cheese. Livestock Sci 139(1–2):91–99

47. Williams AG, Audsley E, Sandars DL (2006) Determining the environmental burdens and resource use in the production of agricultural and horticultural commodities. Main report. Defra Research Project IS0205. Cranfield University and Defra, Bedford

Figure 9, as well as several supplemental files, is not available in this version of the article. To view this additional information, please use the citation on the first page of this chapter.

PART III

MEAT INDUSTRY

CHAPTER 6

A Comparison of the Greenhouse Gas Emissions From the Sheep Industry With Beef Production in Canada

JAMES A. DYER, XAVIER P. C. VERGÉ,
RAYMOND L. DESJARDINS, AND DEVON E. WORTH

6.1 INTRODUCTION

In Canada the main livestock-based industries are beef, pork, dairy, eggs and poultry. In comparison to beef and pork, sheep-based products play only a minor role in the Canadian agri-food sector (Statistics Canada, 2009). Consequently, Canada's contribution to the global market for lamb is very small (Walker, 2008). The small size of the Canadian sheep population, the low volume of exports and the growing global demand for protein suggest an opportunity to expand the sheep industry in Canada. Impending climate change will introduce more variability to food production (Bonatti, Schlindwein, de Vasconcelos, Sieber, & D'Agostini, 2013). It is important for land use policies, including livestock, to be diverse and to leave a minimal carbon footprint. Since both beef cattle and sheep are ruminants, the carbon footprint of sheep production can be defined most effectively by using the beef industry as a benchmark (Edwards-Jones, Plassmann, & Harris, 2009). Whether the current approach to raising sheep in Canada, which is relatively intensive, has a greater GHG emission intensity than beef production needs to be determined. If this was found to be true, then, from the perspective of GHG emissions, the land needed to expand sheep would be better used to support additional beef cattle.

In addition to being the largest livestock industry in Canada, and the largest agricultural emitter of Greenhouse Gases (GHG) (Vergé et al., 2012), beef production is the most GHG-intensive source of protein of Canada's five major livestock industries (Dyer, Desjardins, & Worth, 2010a). Considerable information is available regarding the GHG emissions from the Canadian beef industry (Beauchemin, Janzen, Little, McAllister, & McGinn, 2010; Desjardins et al., 2012; Vergé, Dyer, Desjardins, & Worth, 2008a). Unlike the grain-fed hogs and poultry, the reliance of beef production on roughages makes enteric methane a major term in the GHG emissions budget of beef cattle (Capper, 2012; IPCC, 2006). There is currently no effective mitigation for enteric methane emissions (Cottle, Nolan, & Wiedemann, 2011). However, there is widespread dependence on feedlot operations in both the Canadian and U.S. beef industries, where more grain in the cattle diet reduces the intensity of enteric methane emissions (Capper, 2012; Vergé et al., 2008a).

Quantifying the GHG emissions from the Canadian sheep industry was the main goal of this paper. The other agro-ecological benefits were excluded from this assessment. The GHG emissions budget of sheep was determined using the Unified Livestock Industry and Crop Emissions Estimation System (ULICEES) described by Vergé et al. (2012). Comparing the carbon footprints of the Canadian beef and lamb industries was the second goal of this paper. Since ULICEES has already been applied to the Canadian beef industry, it facilitated a comparison of this industry with the Canadian beef industry on the same computational basis. To verify that ULICEES was an appropriate model for Canadian sheep production, the GHG emissions budget of sheep was compared to several offshore national sheep industry GHG emission estimates.

6.2 BACKGROUND

Although these two grazing animals have fundamentally different mouth parts (OMAFRA, 2011) and differ in their preference for the type of forage they eat (Van Dyne, Brockington, Sxocs, Duek, & Ribic, 1980), they are both ruminants and emit enteric methane. The dryness of sheep manure compared to beef cow manure (ASAE, 2003) is an important difference between these two ruminants because manure storage is the source of two important types of GHG (Vergé et al., 2012). The storage of dry manure produces large emissions of N_2O while storage of liquid manure produces large emissions of CH_4 (Janzen, Desjardins, Rochette, Boehm, & Worth, 2008). Since sheep manure that is not directly deposited in the pasture is managed mainly by dry storage, emissions of N_2O are the main concern.

While the need for domestic wool made sheep an integral part of Canada's early farming history (CCWG, 2011), easier management has made cattle the dominant farm animal on prairie grazing lands (Vergé et al., 2012). The resulting reduction in sheep numbers coincided with the spread of invasive weeds on western rangelands (Walker, Coffey, & Faller, 2006). Since sheep eat weeds such as leafy spurge, which are toxic to cattle, it has been suggested that producers could use these animals for non-chemical weed control (SMA, 2008).

6.3 METHODOLOGY

Since ULICEES has been described in detail elsewhere (Vergé et al., 2012), this section presents only the main concepts used and the adaptations needed for the sheep industry. ULICEES was created by assembling the four groups of livestock-specific GHG computations from the Canadian beef, dairy, pork and poultry industries (Vergé, Dyer, Desjardins, & Worth, 2007, 2008a, 2008b, 2009) in one spreadsheet model. As well as the direct emissions from livestock, these calculations account for GHG emissions from the crop complex that is used to feed each livestock population, and manure characteristics and storage practices of each livestock type. The ULICEES model accounted for the N2O emissions from sheep manure which was mainly handled as dry manure. Because 2001 was the most recent year with livestock diet survey data (Elward, McLaughlin, & Alain, 2003), ULICEES was initially applied to 2001 (Vergé et al., 2012). For this analysis, however, ULICEES was updated to 2006, the closest census year to the mid-point of the 2000-2009 study period used in the sheep report for Canada (Statistics Canada, 2009). This update of ULICEES had to allow for the changes in tillage practices, crop areas, fertilizer applications and livestock populations.

6.3.1 MODELING THE GHG BUDGET OF CANADIAN SHEEP

To take into account the complete lifecycle of meat animals, ULICEES calculates the GHG emissions from all age-gender categories of each type of Livestock. The age-gender categories used in ULICEES for sheep in the present study were mature breeding animals (ewes and rams), breeding (replacement) lambs and market lambs (Statistics Canada, 2003). The livestock GHG assessments include fossil CO_2, CH_4 and N_2O emissions. Emissions of CH_4 and N_2O were distributed over age-gender categories within each livestock type based on differences in feed intake and live weight (LW).

6.3.1.1 METHANE EMISSIONS

Emissions from enteric fermentation and manure storage were calculated on a per-head basis using the ULICEES model (Vergé et al., 2012) which relies on IPCC Tier 2 methodology (IPCC, 2006). Because the two sources require different management, ULICEES quantified manure methane separately from enteric methane. Both types of methane emissions from each age-gender category were multiplied by each respective category population.

6.3.1.2 NITROUS OXIDE EMISSIONS

ULICEES also uses the Tier 2 methodology from IPCC (2006), modified for Canadian conditions by Rochette et al. (2008), to estimate nitrous oxide emissions. For sheep, nitrogen (N) excretion rates were based on Dry Matter Intake (DMI) (Vergé et al., 2012). The Canadian DMI average values (Marinier, Clark, & Wagner-Riddle 2004) were indexed to the average animal weights for the two regions and the age-gender categories before being integrated over the population in each age-gender category.

6.3.1.3 CARBON DIOXIDE EMISSIONS

The fossil CO_2 emissions budget for livestock in ULICEES was based on the six farm energy terms defined by Dyer and Desjardins (2009). Provincial fossil CO_2 emission rate estimates for these six energy terms for 2006 provided by Dyer, Desjardins, McConkey, Kulshreshtha, and Vergé (2013) were incorporated into ULICEES. The farm fieldwork term and fertilizer and machinery supply energy terms were extrapolated from the energy use for all field crops according to Dyer and Desjardins (2009) based on the complex of crop land that supports the sheep industry. The three indirect energy terms, on-farm transport, heating fuels and electricity, could not be related to the livestock crop complexes (Dyer et al., 2013). Instead they were taken from the 1996 Farm Energy Use Survey (FEUS) reported by CAEEDAC (2001). The sheep to beef cattle ratio of total LW was used to scale the FEUS energy quantities for beef to the sheep population because the Canadian sheep industry was too small to be treated as a separate farm type in the FEUS (CAEEDAC, 2001).

6.3.2 SHEEP CROP COMPLEX

The livestock crop complex concept was used in the four previous livestock GHG emission budgets in Canada (Dyer, Vergé, Desjardins, & Worth,

2008; Vergé et al., 2007, 2008a, 2008b, 2009). This concept set the limits of the livestock production systems and has been used to quantify the cropland not used to support livestock in Canada (Dyer, Vergé, Kulshreshtha, Desjardins, & McConkey, 2011). It was used in the present study to describe both the area needed to grow the crops that feed all animals in the sheep and beef industries, as well as the GHG emissions caused by the production of those crops. The crop complex area includes both the roughage and grain crops in the animal diet. A relative comparison of areas in the Sheep Crop Complex (SCC) with the Beef Crop Complex (BCC) on the basis of five crop type groups is shown in Table 1. Corn silage was in a crop category by itself because it is both a roughage and an annual crop. In ULICEES, only improved pasture areas were considered in the GHG emission budgets because unimproved pasture does not receive any farm inputs (Vergé et al., 2012). The crop areas, shown in Table 1 as percent of all areas in each crop complex, were estimated from livestock diets using the ULICEES model.

TABLE 1: Livestock (LS) crop complexes of the sheep and beef industries in Eastern and Western Canada based on five different feed types during 2006

Feed type	Annuals		Alfalfa mixes	Other hay	Improved pasture	All crop complex land
	Grain	Silage				
% of area in the crop complex of each LS type						
Eastern Canada						
sheep	7	2	40	33	18	100
beef	16	3	27	28	25	100
Western Canada						
sheep	12	0	42	13	32	100
beef	21	0	29	14	36	100

6.3.2.1 GRAIN CROP

As was done for the four previous complexes, the grain area in the SCC is the product of the sheep population, their diet and the yield of each feed grain, integrated over all grain crops in the sheep diet. Because Statistics

Canada gives livestock rations for a whole year (Elward et al., 2003) these data give the total quantity of feed consumed during the life of market lambs that are younger than one year when slaughtered. The number of market lambs required to determine the annual GHG emissions budget must represent the average yearly population. Therefore, feed intake over the life of these short-lived animals was expanded to obtain the equivalent feed quantity over all sheep that lived during the full year. Rations were multiplied by the ratio of market lamb population slaughtered in 2006 (Statistics Canada, 2009) to the population of all living market lambs at the time of the 2006 agricultural census. The ratio for Canada was 1.92. This high ratio is a reflection of the lives of market lambs only being about 190 days, compared to a full year.

6.3.2.2 PERENNIAL FORAGES

To estimate the area of perennial forage in the SCC, the provincial crop areas provided by Statistics Canada (2002) had to be partitioned among the dairy, beef and sheep populations. However, because there were no yield data collected for forage crops in Canada, the forage components of the SCC could not be calculated in the same way as the grain components. Since the harvest from this land was limited to fodder for sheep, beef or dairy cattle, the consumption amounts for each province, forage type and age-gender category (Elward et al., 2003) were used to partition the forage areas among these categories over each of the three ruminant livestock types. The four sources of roughage as defined by the Canadian agricultural census (Statistics Canada, 2002) include tame or seeded pasture, alfalfa and alfalfa mixes, corn silage and all other tame hay and fodder crops (Table 1).

6.3.3 ANIMAL POPULATION AND PRODUCTION

6.3.3.1 ANIMAL POPULATIONS

Assessing all animal categories during one year would be equivalent to the complete life cycle of the meat produced in that year (Dyer, Vergé,

Desjardins, Worth, & McConkey, 2010b). Because ULICEES calculates GHG emissions separately for each age-gender category, the true carbon footprint of slaughtered lambs is defined by the GHG emissions from the whole sheep population. The three age-gender categories used to describe the sheep population are presented in Table 2. Rams were combined with ewes as one category because of the very small number of rams.

6.3.3.2 CANADIAN MARKET LAMB PRODUCTION

The GHG emission intensity of Canadian sheep was based on the LW at the time of slaughter, rather than the average LW during its growing life span. Since the ideal market LW is 45 to 50 kg (BCMAL, 2013), 48 kg was used as a benchmark LW in the emission intensity estimates. Due to data limitations and the small size of the industry, we did not try to differentiate east-west market lamb LW. Since sheep farmers must cull their oldest ewes every year, 22% of the breeding ewes contribute to the carcasses going to market (Hale, Coffey, Bartlett, & Ahrens, 2010), equivalent to an average of 4.5 reproductive years for the breeding ewe. In comparison, a six year breeding life was used for beef (Vergé et al., 2008a). The LW assumed for the culled ewes was 57 kg (Hale et al., 2010), while the LWs for slaughtered calves, heifers, steers and culled cows were as used in ULICEES (Vergé et al., 2012).

6.3.3.3 BEEF TO LAMB PROTEIN DIFFERENCES

The difference in GHG emission intensities between the beef industry and the sheep and lamb industry was assessed on the basis of protein supply. In this context, protein is taken to include only human edible protein (excluding blood meal, pet food, edible offal and leather). This comparison does not allow for potential nutritional differences between beef and lamb. The LW conversions to protein were 6.4% for slaughter lambs (Pouliot et al., 2009; USDA, 2009) and 8.3% for slaughter steers and heifers (Dyer et al., 2010a; USDA, 2009).

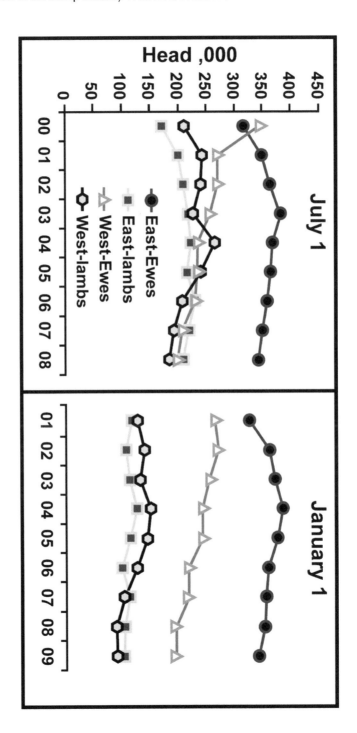

FIGURE 1: Breeding ewe and lamb populations over Eastern and Western Canada recorded on July 1 and January 1 between July, 2000 and January, 2009

TABLE 2: Eastern and Western Canadian sheep populations by age-gender category recorded on July 1 for the 2006 census

Region	Head,000			
	Rams and Ewes	Replacement lambs	Market lambs	Total population
East	369	52	164	584
West	234	41	164	439
Canada	602	93	328	1,023

6.4 RESULTS

6.4.1 LAND AREAS AND SHEEP POPULATIONS

There are noticeable differences in the land use shown in Table 1 between the east and west for both the BCC and SCC. Silage is only fed in measurable quantities in Eastern Canada. The areas in the alfalfa mixes (third column) were almost 50% higher in the SCC than in the BCC in both regions. The harvested perennial forage (columns 3 and 4) accounted for relatively more area in the SCC than in the BCC in both the east and the west. The beef industry in the east used 40% more of its crop complex area as improved pasture land (fifth column) than the sheep industry, whereas this difference was only 11% in the west. Most importantly, in both the SCC and the BCC, Western Canada had appreciably more of its land base in feed grains (first column) than did Eastern Canada.

Table 2 shows that breeding sheep are just over half of the whole Canadian population, which is a reflection of the short lifespan of the slaughter lambs. Although these replacement lambs are not counted as breeding stock, these counts include rams which make up 5% of the ewe population. In the 2006, July 1 census records, market lambs are more than two thirds of the ewe population in Western Canada, but are just under half of the ewe population in Eastern Canada. These lamb-to-ewe ratios are only a little lower than the marketing survey records shown in Figure 1 for July 1, 2006 (Statistics Canada, 2009). In Western Canada, the ewe population declined steadily by about 20% throughout the nine years, while in Eastern Canada, the ewe population did not start to decline until after 2004 (Figure

1). Overall, sheep populations are a little lower in the west than in the east, with a little over a million sheep in Canada (Table 2).

The total GHG emission budgets for sheep and beef cattle are shown in Table 3. The proportion of CH_4, N_2O and fossil CO_2 emissions (in CO_2e) from sheep in Western Canada were 54%, 35% and 11%, respectively, while for beef they were 60%, 28% and 12%, respectively (Table 3). The distribution for beef was similar to the 58%, 33% and 9% distribution of these three GHGs reported for beef on an experimental farm in Alberta between 2008 and 2010 (Basarab et al., 2012). Due to the very small population, around 1 million head in 2006, GHG emissions from raising beef cattle were roughly two orders of magnitude higher than the GHG emissions from raising sheep (Table 3). For beef, the GHG emissions in the west were much higher than in the east, while the east-west distribution of GHG emissions for the sheep industry was consistent with the 60% to 40%, east to west, sheep population split shown in Table 2. In the west the percentages of the total GHG emissions that CH_4 emissions accounted for from the two industries were similar. The CH_4 share of GHG emissions was slightly lower in the east for both industries. For sheep, as well as for eastern beef, N_2O emissions exceeded fossil CO_2 emissions by three and half times.

TABLE 3: Greenhouse gas (GHG) emissions from the sheep and beef industries in Canada in 2006

				Tg CO2e				
	CH_4	N_2O	CO_2	Total GHG	CH_4	N_2O	CO_2	Total GHG
		Sheep				Beef		
East	0.10	0.09	0.03	0.22	3.05	1.87	0.66	5.58
West	0.07	0.05	0.01	0.13	18.10	8.07	1.82	27.99
Canada	0.17	0.14	0.04	0.35	21.15	9.94	2.48	33.57

6.4.2 GHG EMISSION INTENSITY

Table 4 shows the comparison of the sheep and lamb industry with the beef industry on the basis of the GHG emissions (in CO_2e) per weight of protein produced. In terms of protein based intensity, fossil CO_2 and CH_4 emissions from sheep exceeded those from beef cattle by 30%. For the N_2O emission intensity, the lamb intensity was 2.2 times as high as that of beef. For total GHG emission intensities, lamb production exceeded beef by factors of 1.31 in the east, 1.65 in the west and 1.65 for Canada. In both lamb and beef production, CH_4 accounted for 50% to 60% of total emissions, while fossil CO_2 accounted for 7% to 11% of the total and N_2O accounted for 30% to 40% of the total. For lamb production, CH_4 emission intensities as a share of the total GHG emissions were slightly lower than for beef, while the lamb N_2O emission intensity share of its total was slightly higher. These differences reflect the differences in manure since differences between sheep and beef manure affects both CH_4 and N_2O emission rates. The spread between CH_4 and N_2O emission intensities was slightly higher in the west than in the east for both industries. This same pattern was seen in the total GHG emission beef-sheep comparisons (Table 3).

TABLE 4. Protein-based emission Intensity of CH_4, N_2O, fossil CO_2, and total GHG from sheep and beef cattle in Eastern and Western Canada during 2006

| | kg CO_2e/kg protein | | | | | | | |
| | CH_4 | N_2O | CO_2 | Total GHG | CH_4 | N_2O | CO_2 | Total GHG |
			Sheep				Beef	
East	102	93	27	223	93	57	20	170
West	113	73	21	207	81	36	8	126
Canada	107	85	24	217	83	39	10	131

The GHG emission intensity reported in the LCA for sheep in France (Gac, Ledgard, Lorinquer, Boyes, & Le Gall, 2012) without an adjustment for the wool byproduct was 12.9 kg CO_2e/kg LW. Williams, Audsley, & Sandars (2012) found the GHG emission intensity for lamb production in northern England to be 22 tCO_2e/t edible carcass. Using the percents of carcass per LW and of carcass after fat trim presented for beef by Dyer et al. (2010a), this result was converted to 12.3 kg CO_2e/kg LW (Figure 2).

The GHG emission intensity for lamb production was found to be 12.9 kg CO_2e/kg LW in Wales (Edwards-Jones et al., 2009) and 10.0 kg CO_2e/kg LW for the extensive production system in Ireland (Casey & Holden, 2005). The New Zealand (NZ) results reported by Cac et al. (2012) were 8.5 kg CO_2e/kg LW. The much lower intensity was expected for NZ, due to warmer climatic conditions, year-round grazing in perennial pastures and no GHG emissions linked to the production of harvested feed and manure management.

The comparison of this paper with the emission intensities from sheep in other countries required that GHG emission intensities be expressed on a LW, rather than protein, basis. The gas-specific emission factors shown in Table 5 represent close to primary calculations within ULICEES. However, they could not be broken down to the per-number of head basis, because of weight and diet differences among the age-gender categories in each live-stock type. Nor could they be broken down by area because the SCC involves three distinct types of land use (pasture, harvested perennial forage and annual feed crops). These emission factors are the components of the LW-based emission intensities for Canadian lambs shown in Figure 2.

TABLE 5: GHG emission factors for sheep and beef cattle in Eastern and Western Canada during 2006 based intensity of emissions of CH_4, N_2O, and fossil CO_2 per unit of LW[1] of slaughter animals

	kg CO2e/kg LW					
	CH_4	N_2O	CO_2	CH_4	N_2O	CO_2
Region	Sheep			Beef		
East	6.5	6.0	1.7	7.7	4.7	1.7
West	7.3	4.6	1.3	6.7	3.0	0.7

[1]*LW = Live Weight of slaughter animals.*

The average GHG emission intensity of lambs from Table 4 (based on 6.4% protein/LW) was 13.9 kg CO_2e/kg LW for Canada, 14.2 kg CO_2e/kg LW for the east, and 13.2 kg CO_2e/kg LW for the west (Figure 2). The Canada-wide LW based GHG emission intensity for lambs exceeded the English and French LW based emission intensities by 13% and 7%, respectively. The emission intensity of the Canadian sheep industry was almost double that of the New Zealand sheep industry. Using 8.3% protein for slaughter steers and heifers, the average Canadian GHG emission intensity of beef from Table 4 was 11.6 kg CO_2e/kg LW. For the eastern and western beef industries, the intensities were 14.2 kg CO_2e/kg LW and 11.2 kg CO_2e/kg LW, respectively. In their LCA for sheep, Gac et al. (2012) proposed two adjustment factors for the wool byproduct in the carbon footprint of lamb production in France. The reduction factor based on economic allocation was considered negligible, at 0.3%, in France, given the low global value of wool. However, if the reduction factor based on mass allocation in France, 10.4%, was applied to the Canadian lamb industry, the GHG emission intensity of lambs would be reduced to 12.4 kg CO_2e/kg LW for Canada.

6.5 DISCUSSION

The high ratios of ewes and rams to replacement lambs in Table 2 suggest that the reproductive lifespan of a ewe is about 6 years in Canada which is appreciably longer than the 4.3 year lifespan reported by Williams et al. (2012) or the 4.5 years suggested by Hale et al. (2010). Rather than a 6 year breeding life, however, it is more likely that Statistics Canada has miscounted the age-gender categories of sheep. For example, replacement ewe-lambs which have not yet given birth or ewes that are soon destined for slaughter as culls may both have been counted as breeding ewes. The margin by which breeding sheep (rams and ewes) outnumbered the combined replacement and slaughter lambs was 70% in the east and 10% in the west based on the 2006 census data. Also, the ratio of rams and ewes to the replacement lambs in Table 2 is higher in the east. These east-west differences are consistent with Figure 1.

Even though the sheep GHG emission intensities generated by ULICEES (Figure 2) exceeded the French and English sheep industries, these differences were considered to be reasonably close. The higher Canadian GHG intensities are partly due to the more severe climate in Canada. They were also partly due to the intensive production system used in Canada compared to the offshore lamb industries (Edwards-Jones et al., 2009), particularly Ireland. Figure 2 suggests that beef from Eastern Canada has a higher carbon footprint than lamb from the five offshore lamb industries that were compared here. However, due to the differences in protein to LW ratios, emission intensity estimates based on LW mask 33% of the difference between Canadian lamb and beef based on edible protein. Hence, the differences seen in Table 4 give a more realistic comparison of eastern Canadian beef to offshore lamb than shown in Figure 2.

The east-west differences in GHG emissions and emission intensities for both the sheep and beef industries (Tables 3 and 4) were mainly due to higher N_2O emissions related to the more humid conditions and wetter soils in Eastern Canada (Desjardins et al., 2012; Rochette et al., 2008). While they were not allowed for in this analysis, the age and weight of lambs before slaughter, fecundity and breed, may be possible sources of east-west differences. However, with the market for Canadian wool being so low, there is no incentive for Canadian sheep farmers to choose breeds based on wool production rather than meat (Hosford, 2007; Stewart, personal communication).

The sheep and lamb industry in Canada is very small relative to the sheep industries in England and France (Walker, 2008) and in comparison to the Canadian beef industry (Table 2). It is dispersed over large areas dominated by other farm types. Consequently, it is not surprising that the data on sheep breeds and management needed to make in depth regional comparisons have not been gathered in Canada. Being a small industry dispersed over long distances and a range of climates, wide variances should be expected on the mean Canadian GHG emissions presented in this paper.

The agreement with the distribution of emissions over the three GHGs between the estimated western beef industry (Table 3) with the measured emissions from the beef herd on the Alberta experimental farm was reasonably close. This indirect verification for the beef component of ULICEES, even though it was limited to one site, provided some confidence in the value of ULICEES as a tool for this analysis. The east-west difference in the GHG emissions from the beef industry is consistent with GHG emission assessment reported by Desjardins et al. (2012). Whether using rangeland to graze sheep is sustainable would depend on a range of ecological factors as well as GHG emissions, and all of these factors must be considered in evaluating this land use policy.

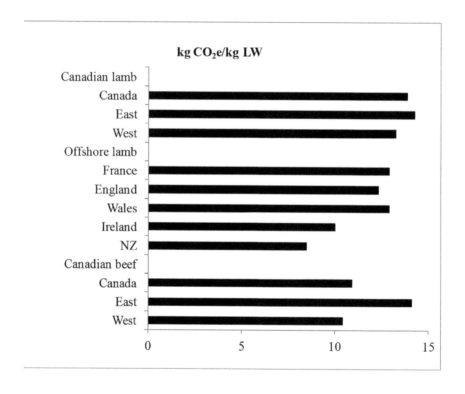

FIGURE 2: Comparisons of the GHG emission intensities of Canadian beef, lamb and offshore lamb, including France, England, Wales, Ireland and New Zealand (NZ), on the basis of live weight (LW) production for 2006

6.6 CONCLUSION

Perhaps the most surprising aspect of the current Canadian sheep industry was its intensive production practices and the relatively high dependence on grains in the sheep diet. With the GHG emission intensity of the current sheep industry exceeding that of beef throughout Canada and for all three GHGs (Tables 4 and 5), there appears to be little benefit in expanding the sheep industry under the current grain dependent production system in either Eastern or Western Canada, at least from the perspective of the livestock carbon footprint. Differences in the foraging habits of sheep and cattle may mean some ecological benefits in grazing sheep that should be explored. Although the protein basis of comparison was more dramatic, both the LW and protein based GHG emission intensity indicators show the disadvantage of relying on lamb as a protein source, compared to beef.

For more accurate assessment of the role that sheep should play in the Canadian livestock industries, a repeat of the 2001 diet survey of all livestock types (Elward et al., 2003) is needed. The increasing availability of biofuel feedstock byproducts can do much to increase dietary energy and protein and reduce the carbon footprints of all livestock commodities in Canada. This new form of land use adds further incentive to update the livestock diet information base across Canada.

The next step following the research recommended above will be an LCA which would include all of the processing of the meat (lamb), offal, hide and wool products and byproducts. While ULICEES is not an LCA model, it has been interfaced with an LCA for the carbon footprint of the dairy industry (Vergé et al., 2013) and has been applied to an LCA of the vertical integration of the beef industry (Desjardins et al., 2012). Although they were beyond the scope of this paper, ecological and social values need to be integrated with the GHG emission estimates in this LCA. The ULICEES model and the results shown in this paper would provide a basis for an LCA of Canadian lamb production. As the field monitoring research and data upgrades proposed above progress, sensitivity analysis should be done on the ranges in the terms that go into the ULICEES estimates of the GHG emissions from sheep and lambs in Canada.

While both economic and mass allocation would have to be investigated for dealing with wool on an international scale, the very low economic value of this byproduct in France (Gac et al., 2012) and the growing preference for breeds that shed their fleece (Walker, 2008) effectively eliminates wool from the carbon footprint of the lamb industry, at least in Canada. Since the market demand for wool is almost gone, economics was the more meaningful byproduct allocation for both the Canadian and French sheep industries. A more important justification than wool for continued support of lamb production in Canada would be the ability of sheep to digest plant types that are unsuitable as cattle feed, and in some cases (such as leafy spurge) are toxic to cattle. The role that sheep in concert with cattle could play in making sustainable use of rangeland, where a range of plant species must be allowed to grow, should be investigated as a land use policy in Western Canada. This policy, however, would be a departure from the current intensive approach to lamb production. Whether using rangeland to graze sheep is sustainable would depend on a range of ecological factors as well as GHG emissions, and all of these factors must be considered in evaluating this land use policy.

REFERENCES

1. ASAE. (2003). Manure Production and Characteristics - D384.1 FEB03- 4p. American Society of Agricultural and Biological Engineers (ASAE), St. Joseph, MI, USA.
2. Basarab, J., Baron, V., López-Campos, Ó., Aalhus, J., Haugen-Kozyra, K., & Okine, E. (2012). Greenhouse Gas Emissions from Calf- and Yearling-Fed Beef Production Systems, With and Without the Use of Growth Promotants. Animals, 2, 195-220. http://dx.doi.org/10.3390/ani2020195
3. BCMAL. (2013). Sheep - How are sheep produced? British Columbia Ministry of Agriculture (BCMAL). Province of British Columbia. Retrieved from http://www.agf.gov.bc.ca/aboutind/products/livestck/sheep.htm
4. Beauchemin, K. A., Janzen, H. H., Little, S. M., McAllister, T. A., & McGinn, S. M. (2010). Life cycle assessment of greenhouse gas emissions from beef production in western Canada: A case study. Agricultural Systems, 103, 371-379. http://dx.doi.org/10.1016/j.agsy.2010.03.008
5. Bonatti, M., Schlindwein, S. L., de Vasconcelos, A. C. F., Sieber, S., D'Agostini, L. R., Lana, M. A., … Canci, A. (2013). Social organization and agricultural strategies to face climate variability: A case study in Guaraciaba, Southern Brazil. Sustainable Agriculture Research, 2(3), 118-125. http://dx.doi.org/10.5539/sar.v2n3p118

6. CAEEDAC. (2001). A Review of the 1996 Farm Energy Use Survey (FEUS). A Report to Natural Resources Canada (NRCan) by The Canadian Agricultural Energy End-Use Data Analysis Centre (CAEEDAC). Retrieved from http://www.usask.ca/agriculture/caedac/pubs/pindex.htm

7. Casey, J. W., & Holden, N. M. (2005). Holistic analysis of GHG emissions from Irish livestock production systems. Paper No. 054036. St. Joseph, MI, American Society of Agricultural and Biological Engineers. Retrieved from https://elibrary.asabe.org/azdez.asp?AID=19478&T=2

8. Capper, J. L. (2012). Is the Grass Always Greener? Comparing the environmental impact of conventional, natural and grass-fed beef production systems. Animals, 2, 127-143. http://dx.doi.org/10.3390/ani2020127

9. CCWG. (2011). Canada's wool to market. Canadian Cooperative Wool Growers Ltd. (CCWG). Retrieved from http://www.wool.ca/about_wool

10. Cottle, D. J., Nolan, J. V., & Wiedemann, S. G. (2011). Ruminant enteric methane mitigation: a review. Animal Production Science, 51(6), 491-514. http://dx.doi.org/10.1071/AN10163

11. Desjardins, R. L., Worth, D. E., Vergé, X. P. C., Maxime, D., Dyer, J., & Cerkowniak, D. (2012). Carbon Footprint of Beef Cattle. Sustainability, 4, 3279-3301. http://dx.doi.org/10.3390/su4123279

12. Dyer, J. A., & Desjardins, R. L. (2009). A review and evaluation of fossil energy and carbon dioxide emissions in Canadian agriculture. J. Sustain. Agric., 33(2), 210-228. http://dx.doi.org/10.1080/10440040802660137

13. Dyer, J. A., Desjardins, R. L., McConkey, B. G., Kulshreshtha, S., & Vergé, X. P. C. (2013). Integration of farm fossil fuel use with local scale assessments of biofuel feedstock production in Canada. Chapter 4; In Biofuel Economy, Environment and Sustainability (pp. 97-122). Zhen Fang (Ed.) InTech Open Access Publisher. Rijeka, Croatia.

14. Dyer, J. A., Vergé, X. P. C., Desjardins, R. L., & Worth, D. (2008). Long-term trends in the greenhouse gas emissions from the Canadian dairy industry. Can. J. Soil Sci., 88, 629-639. http://dx.doi.org/10.4141/CJSS07042

15. Dyer, J. A., Vergé, X. P. C., Desjardins, R. L., & Worth, D. E. (2010a). The Protein-based GHG Emission Intensity for Livestock Products in Canada, J. Sustain. Agric., 34(6), 618-629. http://dx.doi.org/10.1080/10440046.2010.493376

16. Dyer, J. A., Vergé, X. P. C., Desjardins, R. L., Worth, D. E., & McConkey, B. G. (2010b). Understanding, Quantifying and Reporting Greenhouse Gas Emissions from Canadian Farmland. Sustainable Futures: The official publication of the Agricultural Institute of Canada, (pp. 10, 12).

17. Dyer, J. A., Vergé, X. P. C., Kulshreshtha, S. N., Desjardins, R. L., & McConkey, B. G. (2011). Residual crop areas and greenhouse gas emissions from feed and fodder crops that were not used in Canadian livestock production in 2001. J. Sustain. Agric., 35(7), 780-803. http://dx.doi.org/10.1080/10440046.2011.606493

18. Edwards-Jones, G., Plassmann, K., & Harris, I. M. (2009). Carbon footprinting of lamb and beef production systems: insights from an empirical analysis of farms in Wales, UK. Journal of Agricultural Science, 147, 707-719. http://dx.doi.org/10.1017/S0021859609990165

19. Elward, M., McLaughlin, B., & Alain, B. (2003). Livestock Feed Requirements Study 1999-2001. Catalogue No. 23-501-XIE, Statistics Canada, p. 84.

20. Gac, A., Ledgard, S., Lorinquer, E., Boyes, M., & Le Gall, A. (2012). Carbon footprint of sheep farms in France and New Zealand and methodology analysis. In 8th International Conference on LCA in the Agri-food Sector, Parallel Session 3C: Sheep and Dairy Production Systems. Saint Malo, France.

21. Hale, M., Coffey, L., Bartlett, A., & Ahrens, C. (2010). Sustainable and Organic Production, A project of the National Center for Appropriate Technology (NCAT), National Sustainable Agriculture Information Service. Retrieved from http://www.cultivatingsuccess.org/instructors/SSFR%20Readings%202011/9.5_Sust_Sheep.pdf

22. Hosford, S. (2007). Wool Marketing in Canada. Agriculture and Rural Development (ARD) of Alberta. Retrieved from http://www1.agric.gov.ab.ca/$department/deptdocs.nsf/all/sis11784

23. IPCC. (2006). Intergovernmental Panel on Climate Change. Guidelines for National Greenhouse Gas Inventories. Volume 4: Agriculture, Forestry and Other Land Use. Chapter 10: Emissions from Livestock and Manure Management. Retrived from http://www.ipcc-nggip.iges.or.jp/public/2006gl/index.htm

24. Janzen, H. H., Desjardins, R. L., Rochette, P., Boehm, M., & Worth, D. (Eds). (2008). Better Farming Better Air. Agriculture and Agri-food Canada (p. 151). Ottawa, Ontario, Canada.

25. Marinier, M., Clark, K., & Wagner-Riddle, C. (2004). Improving estimates of methane emissions associated with animal waste management systems in Canada by adopting an IPCC Tier 2 methodology. Technical Report, p. 30. University of Guelph.

26. OMAFRA. (2011). Grazing Management: Animals as Grazers, Publication 19, Pasture Production, Chapter 4; Ontario Ministry of Agriculture, Food and Rural Affairs (OMAFRA). Retrieved from http://www.omafra.gov.on.ca/english/crops/pub19/4grazers.htm

27. Pouliot, E., Garie, C., Theriault, M., Avezard, C., Fortin, J., & Castonguay, F. W. (2009). Growth performance, carcass traits and meat quality of heavy lambs reared in a warm or cold environment during winter. Can. J. Anl. Sci., 229-239.

28. Rochette, P., Worth, D. E., Lemke, R. L., McConkey, B. G., Pennock, D. J., Wagner-Riddle, C., & Desjardins, R. L. (2008). Estimation of N2O emissions from agricultural soils in Canada – Development of a country-specific methodology. Can. J. Soil Sci. (Special edition), 88, 641-654. http://dx.doi.org/10.4141/CJSS07301

29. SMA. (2008). Reducing Leafy Spurge's Impact By Using Sheep and Goats, Government of Saskatchewan, Saskatchewan Ministry of Agriculture (SMA). Retrieved from http://www.agriculture.gov.sk.ca/Default.aspx?DN=7984b1a8-5088-4068-a7d4-024f5837e240, Accessed 10 May 2013.

30. Statistics Canada. (2002). Agricultural Census 2001, Cat. No 95F0301XIE.

31. Statistics Canada. (2003). Livestock feed requirements study, 1999-2001, Cat. No 23-501-XIE.

32. Statistics Canada. (2009). Sheep Statistics, vol. 8 no. 1, Cat. No. 23-011-X. p. 30.

33. Stewart, D. W. personal communication. worked as an research scientist in the Research Branch of Agriculture and Agri-food Canada from 1977 to 2002; now a sheep farmer in Eastern Ontario.

34. USDA. (2009). National Nutrient Database for Standard Reference, Release 21 Protein (g) Content of Selected Foods per Common Measure, sorted alphabetically - p. 25 USDA. Retrieved from http://www.ars.usda.gov/Services/docs.htm?docid=8964

35. Van Dyne, G. M., Brockington, N. R., Sxocs, Z., Duek, J., & Ribic, C. A. (1980). Large herbivore sub-systems. In In A. I. Breymeyer & G. M. Van Dyne (Eds.) Grasslands, systems analysis and man. (pp. 269-537). Cambridge, England: Cambridge University Press.

36. Vergé, X. P. C., Dyer, J. A., Desjardins, R. L., & Worth, D. (2007). Greenhouse gas emissions from the Canadian dairy industry during 2001. Agricult. Sys., 94(3), 683-693.

37. Vergé, X. P. C., Dyer, J. A., Desjardins, R. L., & Worth, D. (2008a). Greenhouse gas emissions from the Canadian beef industry. Agricult. Sys., 98(2), 126-134.

38. Vergé, X. P. C., Dyer, J. A., Desjardins, R. L., & Worth, D. (2008b) Greenhouse gas emissions from the Canadian pork industry. Livest. Sci., 121, 92-101. http://dx.doi.org/10.1016/j.livsci.2008.05.022

39. Vergé, X. P. C., Dyer, J. A., Desjardins, R. L., & Worth, D. (2009). Long-term trends in greenhouse gas emissions from the Canadian poultry industry. J. appl. Poult. Res., 18, 210-222. http://dx.doi.org/10.3382/japr.2008-00091

40. Vergé, X. P. C., Dyer, J. A., Worth, D. E., Smith, W. N., Desjardins, R. L., & McConkey, B. G. (2012). A greenhouse gas and soil carbon model for estimating the carbon footprint of livestock production in Canada. Animals, 2, 437-454. http://dx.doi.org/10.3390/ani2030437

41. Vergé, X. P. C., Maxime, D., Dyer, J. A., Desjardins, R. L., Arcand, Y., & Vanderzaag, A. (2013). Carbon footprint of Canadian dairy products Calculations and issues. Journal of Dairy Science, 96(9), 6091-6104. http://dx.doi.org/10.3168/jds.2013-6563

42. Walker, C. (2008). Easier managed sheep and beef cattle; simplified, profitable and productive sheep and beef

43. farming. A Nuffield Farming Scholarships Trust Report. Sponsored by the Royal Highland Agricultural

44. Society of Scotland and the Royal Smithfield Club. Retrieved from

45. http://www.nuffieldinternational.org/rep_pdf/1253801058C_H_Walker_Nuffield_Report_read_only.pdf

46. Walker, J. W., Coffey, L., & Faller, T. (2006). Improving Grazing Lands with Multi-Species Grazing. In A. Peischel & D. D. Jr. Henry (Ed.). Targeted Grazing: A Natural Approach to Vegetation Management and Landscape Enhancement (pp. 50-55). Retrived from www.cnr.uidaho.edu/rx-grazing/Handbook.htm

47. Williams, A., Audsley, E., & Sandars, D. (2012). A systems-LCA model of the stratified UK sheep industry. In 8th International Conference on LCA in the Agri-food Sector, Parallel Session 3C: Seep and Dairy Production Systems. Saint Malo, France.

PART IV

COFFEE AND TEA INDUSTRIES

CHAPTER 7

Tensions Between Firm Size and Sustainability Goals: Fair Trade Coffee in the United States

PHILIP H. HOWARD AND DANIEL JAFFEE

7.1 INTRODUCTION

Addressing sustainability goals has proven to be a successful marketing strategy for firms, both to gain market share and to increase profitability [1]. The wide success of third-party certified ecolabels, such as fair trade and organic, as well as first party sustainability claims, has encouraged more firms, and an increasing number of industries, to develop such efforts [2,3]. These trends have typically been initiated by what Hockerts and Wüstenhagen term "emerging Davids," or smaller, mission-driven firms, but are increasingly attracting "greening Goliaths," or larger, more profit-driven firms [4]. The latter may have less commitment to sustainability ideals (environmental, economic and social), as there are inherent tensions between the missions of firms that are already large (relative

Tensions Between Firm Size and Sustainability Goals: Fair Trade Coffee in the United States.
© *Howard PH and Jaffee D.* Sustainability *5,1 (2013), doi:10.3390/su5010072. Licensed under Creative Commons Attribution 3.0 Unported License, http://creativecommons.org/licenses/by/3.0/.*

to competitors in the industry) and the goal of addressing the industry's negative social and environmental externalities that initially motivated the entry of mission-driven firms [4,5,6].

Not all companies, however, fall neatly into the ideal types of smaller, more committed firms and larger, less committed firms. Most studies have focused on these two categories, without examining large firms that are more committed than typical, or firms with high sustainability commitments that are larger than typical [7,8,9]. Studying the strategies of these businesses can provide a better understanding of the tensions between firm size and sustainability goals, as well as inform efforts to (a) increase the sustainability commitments of large firms, and to (b) assist small firms to scale up while maintaining high commitments to sustainability [4].

We explore these issues through a case study of firms in the U.S. fair trade coffee market using methods of data visualization and content analysis of media coverage. The fair trade market utilizes third-party certification to signify products that meet multiple (social, economic and environmental) sustainability criteria. This case is particularly interesting because of controversial recent policy changes by the entity that was until recently the sole fair trade certifier in the United States. In September 2011 Fair Trade USA (FTUSA, formerly Transfair USA) announced that it was resigning from the international fair trade certification body, Fair Trade International (FTI), and that it would begin certifying coffee produced on plantations, rather than only coffee from small farmer cooperatives, as stipulated by FTI standards. This decision has generated strong disagreements among U.S.-based coffee firms regarding the direction the fair trade system should take. These debates have become increasingly public over the past year [10,11,12,13,14], and have rendered these firms' strategic differences more visible than they were previously.

Below we review literature related to firm size and sustainability efforts, as well as studies that address these issues for fair trade specifically. We then describe how we approached the following two questions, and our findings for each:

- How is the U.S. fair trade coffee industry structured with respect to firm size and degree of engagement with fair trade?
- How have the firms with the highest volumes of fair trade purchases responded to tensions between size and sustainability goals?

The subsequent section discusses our research findings. Starbucks, Green Mountain Coffee Roasters and Equal Exchange were the three firms with the highest volumes of fair trade coffee purchases in the U.S. in recent years, and are therefore arguably contributing substantially to achieving both social and environmental sustainability goals in the fair trade coffee market. However, these three firms are also taking quite divergent paths, as reflected in the levels of fair trade certified coffee purchases relative to their overall coffee volumes, as well as their responses to the changing U.S. fair trade standards. We conclude that among these firms there is an inverse relationship between size and commitment to sustainability practices. In addition, the larger, more profit-driven firms are much less likely to acknowledge tensions between size and sustainability in their public discourse.

7.2 FIRM SIZE AND SUSTAINABILITY GOALS

The positive impacts of large firms participating in sustainability efforts may include (a) an increasing proportion of industry activities that meet sustainability criteria [6,15], (b) an increased availability of products and services for consumers that meet these criteria [16], and/or (c) a greater level of resources devoted to raising consumer awareness of sustainability goals [17]. There are also potential negative impacts, including tendencies for large firms that do participate in sustainability efforts to (a) only apply those practices to a very small percentage of their purchases or sales, while reaping benefits of positive publicity, (e.g., "greenwashing" or "fairwashing") [15,18,19,20] and/or (b) attempt to weaken the original sustainability goals on an industry-wide basis in order to increase profits [9,21].

Conversely, small firms that develop sustainability innovations may have little influence on the industry as a whole unless these innovations spread to larger firms (typically the faster approach), or are scaled up to include a larger proportion of industry practices (typically the slower approach) [4]. This scaling up may involve the growth of small firms into much larger firms, or alternatively, the successful entrance of numerous other small, mission-driven firms into the industry. In any case, such dramatic departures from the status quo require negotiating tensions between the scale of the current market and the institutionalization of more socially and environmentally sustainable practices.

7.2.1 TENSIONS

These size/sustainability tensions have been described in relation to concepts in the economics and business literatures that include externalities, supply chain management, and innovation. Negative externalities are a form of market failure that allow the privatization of profits and the socialization of costs or harms, with the classic example being pollution from manufacturing industries [22]. Moving towards genuine sustainability requires firms and industries to internalize costs that were previously borne by society or the natural environment. Larger firms may find it more difficult to voluntarily reduce their externalities because they have greater investments in practices that shift these costs outside of the firm [4]. Importantly, this includes the shaping of consumer demand so that it is congruent with the continued profitability of big business [23]. Sustainability marketing at this scale therefore requires the expensive and difficult task of reshaping this demand for the mass market, not merely a small niche of consumers [6]. This presents a more significant barrier for small firms attempting to expand their market, because they typically lack the resources to market and advertise to such a broad extent.

Sustainability efforts also frequently require developing new supply chains to source more socially or environmentally sustainable inputs, which can be more difficult for larger firms that deal in high volumes [24]. One issue may simply be a lack of adequate supply of such inputs, while another may be that obtaining these inputs involves high transaction and search costs, as dealing with a large number of suppliers (or customers) is more costly in terms of time and resources than dealing with a smaller number [25]. Firms have an economic incentive to reduce the number of transactions, but the larger the firm relative to the scale of suppliers, the more difficult this becomes. Smaller firms encounter challenges at the distribution end of the supply chain when attempting to scale up, if they are not large enough to overcome these issues for large distributors or retailers [26,27,28].

Sustainability marketing can be viewed as a particular kind of innovation: there is a sizeable literature that suggests larger firms are typically more resistant to innovation, particularly when it involves more than
merely small, incremental changes [4,6,29]. However, not all studies suggest inherent tensions between increased size and sustainability innovations. Many note that economies of scale or greater resources can make
it easier for larger firms to adopt certain types of practices, such as using
more sustainable technologies or requiring their suppliers to conform to
sustainability standards, especially in industries that are capital intensive
or highly concentrated [30,31,32].

7.2.2 TENSIONS IN FAIR TRADE

Fair trade is a certification system that addresses multiple sustainability
goals for primarily agricultural (but also non-food) products, largely produced in the global South. The key criteria underlying fair trade certification include minimum or base prices for producers (or working conditions
for waged laborers), an added development premium, pre-harvest financing, and environmentally sustainable production practices, among others
[19]. Fair trade is the primary product certification that emphasizes the
social conditions of production, and thus places more emphasis on socio-
economic sustainability than other ecolabels, such as organic. Fair trade
has achieved tremendous market success, surpassing $6 billion in global
sales in 2011 with double-digit growth rates being sustained for more than
a decade [33,34]. Studies suggest that this label enhances consumers' positive perceptions of products, although willingness to pay a price premium
is currently limited to a smaller subset of these consumers [35,36,37].

    Tensions between firm size and sustainability have been illuminated
by ongoing controversies over changing fair trade standards, and the increased certification of plantation products and waged labor in the fair
trade system. The moves since 2005 by the global fair trade certifier Fair
Trade International to actively expand the role of plantations for many
crops including tea, fruit, spices, cut flowers, and other products (but not
coffee, cocoa, honey or cotton) have unleashed heated debate about the
movement's historical commitment to small farmers and smallholder pro-

duction [38]. A key concern is that the economic advantages of planta-tions will undercut the small-scale producers who had long been excluded from conventional markets, a context that motivated the creation of the fair trade system orginally [39]. Some researchers have reported negative impacts of including plantations in the fair trade system, including that it undermines stronger state regulation of labor conditions [40], and that in practice it fails to reach the most disadvantaged plantations and their laborers because it privileges higher quality products, which are typically sourced from plantations with better off workers [41]. A study of planta-tion cut flowers in Ecuador, however, found that fair trade plantations had better outcomes for environmental (e.g., reduced pesticide use) and social (e.g. funding for community programs) issues when compared to other plantations [42].

For coffee, plantations may also cause greater negative environmental impacts than smallholder plots, as the former tend to have lower levels of shade tree cover (and thus less biodiversity) and higher use of agrochemi-cals [43,44]. Increased incomes from fair trade have also provided incen-tives for smallholders to maintain high levels of shade, increase tree bio-diversity, and adopt more ecologically sustainable farming practices, such as composting and planting nitrogen-fixing cover crops [19,45]. These practices are frequently transferred to subsistence food plots as well [19].

One example of the tensions between different firm types in the realm of standards involves the criterion requiring that affordable financing (of 60% of the total contract price) be made available to farmer organizations in advance of the harvest, to avoid the classic "debt trap" plaguing small farmers worldwide. This requirement, however, is not honored consistent-ly by firms purchasing fair trade certified coffee. Raynolds reported that profit-driven firms rarely provide advance financing for coffee contracts, while mission-driven firms consistently pre-finance [46]. Another area of-ten honored in the breach is fair trade's stipulation of long-term trading relationships. Raynolds also found that larger, profit-driven firms prefer to enter mainly into one-year (or shorter) contracts with fair trade producer groups, rather than form stable relationships with suppliers as the mission-driven firms do consistently [42,46].

While tensions between size and sustainability goals have been ex-plored for some specific cases in the fair trade system, as described above, there has been insufficient research focusing on the firms making the largest contributions to the industry. To address this gap in the literature we focused on fair trade coffee in the U.S. Our first objective was to characterize the structure of this industry with respect to the size of the firms making fair trade purchases and their degree of engagement with fair trade. Our second objective was to analyze how the firms with the highest volumes of fair trade purchases have responded to tensions between size and sustainability goals, with a focus on recent standards changes.

7.3 HOW IS THE U.S. FAIR TRADE COFFEE INDUSTRY STRUCTURED WITH RESPECT TO FIRM SIZE AND ENGAGEMENT WITH FAIR TRADE?

7.3.1 METHODS

To characterize the structure of the U.S. fair trade coffee industry, we first collected data on firms' total purchases of green (unroasted) coffee, and their purchases of fair trade certified green coffee. The data were compiled from multiple sources, including corporate social responsibility reports; reports by NGOs, academic institutions, and industry; firms' responses to email inquiries; and media reports. While we were not able to obtain data for every firm in the industry, as many choose to keep such data confidential, all of the largest firms in the fair trade market, as well as a sample of smaller roasters, are represented. We constructed several visualizations to aid in data analysis. Compared to text-based formats, this approach permits the communication of large amounts of information more quickly, and with fewer burdens on short-term memory [47,48]. Figure 1 establishes the global context, while the remaining figures explore the U.S. fair trade coffee market in more depth. In Figure 1, proportional circles represent the volume of the total green coffee purchases of the top ten world coffee roasters. This graphic also uses proportional circles to represent the percentage and volume of fair trade certified purchases.

Figure 2 is a multi-variable plot of selected firms engaged in fair trade purchases in the United States. We plotted year of entry into fair trade on

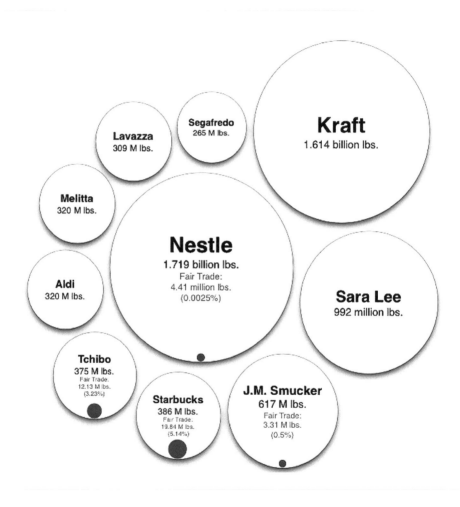

FIGURE 1: Top Ten World Coffee Roasters: Fair Trade Certified and Non-Fair Trade Purchases, 2008 [49].

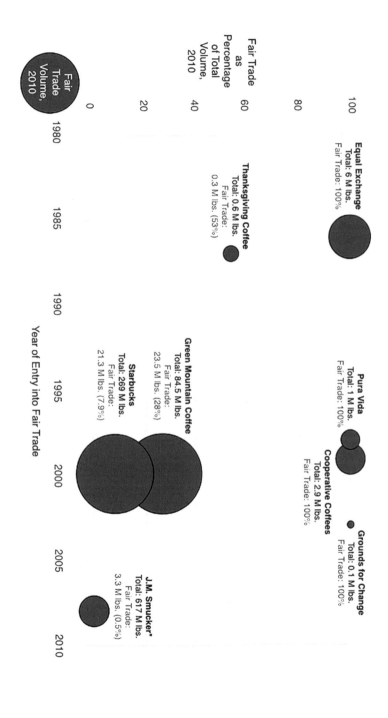

FIGURE 2: Plot of Fair Trade Variables for Selected Coffee Firms in the United States: Year of Entry, Volume, and Percentage of Total Volume [49,50,51,52].

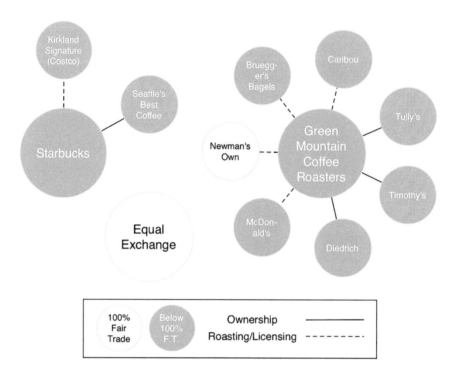

FIGURE 3: Brands and Connections for Firms Accounting for the Highest Proportion of Fair Trade Purchases.

the x axis, fair trade as a percentage of total volume on the y axis, and fair trade volume in the year 2010 as proportional circles. Based on this figure we selected three U.S. firms that accounted for the largest volume of fair trade purchases, while representing both mission-driven (early entry; higher percentage of fair trade) and profit-driven (later entry; lower percentage of fair trade) firms. We then compiled additional information from company websites and scholarly literature to contextualize these firms, and to inform an additional graphic (Figure 3) illustrating the brands that are either owned by, or involved in roasting and/or marketing associations with these firms.

7.3.2 RESULTS

7.3.2.1 THE GLOBAL CONTEXT

Figure 1 depicts the overall context of non-fair trade and fair trade coffee purchases among the top ten firms internationally in 2008. It shows that only four of the ten purchased any fair trade certified coffee. These data suggest an inverse relationship between firm size and commitment to fair trade for those who choose to participate. Fifth-ranked Starbucks (U.S.) was the world's largest purchaser of fair trade certified coffee at this time (nearly 20 million pounds, or 5 percent of its total supply). Sixth-ranked Tchibo (Japan) had a slightly lower commitment to fair trade both by volume and percentage (approximately 12 million pounds; 3 percent of total volume), but this was significantly more than the two largest firms. For no. 1 Nestlé (Switzerland) and no. 4 J.M. Smucker (U.S.), fair trade purchases were relatively similar by volume (approximately 4 and 3 million pounds, respectively), but Nestle had by far the lowest percentage of fair trade purchases (0.0025 percent).

7.3.2.2 THE U.S. CONTEXT

Figure 2 depicts the context in the United States for selected firms that purchased and sold fair trade certified coffee, as of 2010, on three dimensions:

year of entry into fair trade certification, volume of fair trade purchases, and the percentage of total volume those fair trade purchases represent. Earlier entrants are on the left side of the graphic and later entrants are on the right side. Firms that purchase 100% fair trade coffee are at the top and firms with the lowest percentage of fair trade relative to total purchases are at the bottom. The later entrants with lower percentages of fair trade purchases are all large, publicly traded corporations (Green Mountain Coffee Roasters, Starbucks and J.M. Smucker), and the early entrants are smaller firms with much higher percentages of fair trade purchases. Thanksgiving Coffee is somewhat of an anomaly because although an earlier entrant, it does not purchase exclusively fair trade coffee. The remaining smaller firms shown are also all at the 100% level, despite entering the fair trade system in later years.

The firms of greatest interest with respect to our research objectives, however, are those that are making disproportionate contributions to sustainability goals through greater volumes of fair trade purchases (represented by the largest circles in the graphic). In this case Equal Exchange is purchasing more fair trade coffee than other highly committed firms, approximately twice as much as the next largest 100% fair trade firm, Cooperative Coffees (which is a cooperative composed of 23 smaller firms that purchase collectively). At the other end of the spectrum, Green Mountain Coffee Roasters (GMCR) and Starbucks are the two largest buyers of fair trade coffee beans in the U.S. (together accounting for more than 40% of the total volume certified by Fair Trade USA), despite their lower percentages of fair trade relative to total volume [34]. GMCR's fair trade certified purchases in 2010 totaled 23.5 million pounds, which was more than three times as large as Equal Exchange's fair trade purchases, and slightly more than Starbucks' 21 million pounds. GMCR's percentage of fair trade purchases was 27.8%, while Starbucks' percentage was 7.9%. Starbucks' much greater marketing emphasis on its first party (i.e., self-certified) Coffee and Farmer Equity (C.A.F.E.) Practices program, and its low percentage of fair trade purchases, have been characterized by some critics as "dabbling"[7]. Nevertheless, Starbucks was the world's largest buyer of fair trade coffee until 2010, and its size gave the firm a strong voice in the policy direction of fair trade in the U.S., as we discuss further below.

7.3.2.3 CHARACTERISTICS OF FIRMS WITH HIGHEST PROPORTION OF FAIR TRADE PURCHASES

Equal Exchange was the first fair trade coffee company in the U.S., established in 1986. It is organized as a worker-owned cooperative, and purchases coffee from small farmer cooperatives. Its legal structure is unique: if the company is sold, the proceeds must go to another fair trade organization, not to the worker-owners. It is located in West Bridgewater, Massachusetts. Equal Exchange focused exclusively on coffee in its early years, but has since expanded to chocolate, tea, bananas and other products. The firm reported $46.8 million in annual revenue in 2011. Like other 100% fair trade firms, Equal Exchange's marketing efforts are focused primarily on the social responsibility dimension, while the two firms discussed below place far greater emphasis on the dimension of taste [53].

Green Mountain Coffee Roasters originated as a café, but became a publicly traded corporation in 1993. It is headquartered in Waterbury, Vermont. Like the other two firms, GMCR is a specialty coffee roaster, meaning that it focuses on higher quality Arabica coffee varieties that are differentiated from lower-cost, mass-market coffee. The firm began selling fair trade coffee in 2000 after being approached by Transfair USA (now FTUSA), and expanded these sales through a licensing agreement with Newman's Own Organics, a 100% fair trade brand, in 2003 [54]. GMCR also uses other ecolabels, including organic and Rainforest Alliance certifications. The firm reported $2.6 billion in annual revenue in 2011, and has grown very rapidly in recent years due to the popularity of the Keurig single-cup brewing machine (GMCR fully acquired Keurig in 2006) and its patented coffee pods that are inserted in the machine.

Starbucks is the largest coffeehouse chain in the world, and the largest specialty coffee firm. Like GMCR, it is a publicly traded corporation. It is based in Seattle, Washington. The firm began selling fair trade coffee in 2000 as a result of pressure from activist organizations, including Global Exchange, but has received frequent criticism since then for its relatively low percentage of fair trade purchases relative to total volume [19]. The firm reported $11.7 billion in annual revenue in 2011.

Figure 3 depicts the other brands that have connections to these three firms. Starbucks owns Seattle's Best Coffee, and also roasts for the warehouse club Costco's private label brand, Kirkland Signature. Green Mountain Coffee Roasters owns the brands Tully's, Timothy's and Diedrich. In addition the company roasts for McDonald's, Bruegger's Bagels and Caribou Coffee, as well as the 100% fair trade brand Newman's Own. Equal Exchange does not have any subsidiary brands or roasting/licensing arrangements in the U.S., but does have joint branding alliances in other countries [55].

7.4 HOW HAVE THE LARGEST U.S. FAIR TRADE COFFEE FIRMS RESPONDED TO SIZE/SUSTAINABILITY TENSIONS?

7.4.1 METHODS

To characterize the responses of Starbucks, Green Mountain Coffee Roasters and Equal Exchange with respect to sustainability goals and tensions with firm size, we took two approaches. We first visualized and analyzed coffee purchasing behaviors since 2004, to more closely examine firms' levels of commitment over time. Second, we conducted a content analysis of firm responses and the public discourse surrounding Fair Trade USA's exit from the international certification system, which was announced in 2011.

As in the previous section, we collected data for fair trade volume and fair trade coffee as a percentage of total volume from the firms' annual reports, corporate social responsibility (CSR) reports and responses to email queries, as well as company websites. We constructed graphics to depict purchases for each firm from 2004 to 2011 and to compare these over time. Time is plotted on the x axis and volume (total and fair trade certified) on the y axis.

To analyze the discourse of recent changes in the U.S. fair trade system we conducted a search of the LexisNexis news database using the keywords "Fair Trade USA" for All News (English) from September 15, 2011 (the date of Fair Trade USA's announcement of its departure from the FTI system) to October 5, 2012. This resulted in 204 documents, including

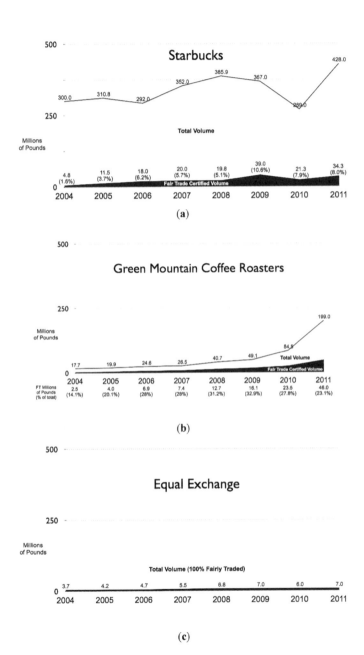

FIGURE 4: Total and Fair Trade Certified Coffee Purchases by Selected Firms, 2004 to 2011. (a) Starbucks [56]. (b) Green Mountain Coffee Roasters [57]. (c) Equal Exchange [58].

newspaper articles, press releases, web blogs, and TV/radio transcripts. This number was reduced to 153 after removing 51 duplicates (i.e., the same article reprinted in multiple outlets). We also searched firm blogs with same search term, resulting in additional 14 documents from Equal Exchange, three documents from GMCR and none from Starbucks. Finally, we included two additional documents from alternative media sources that were referenced by these firms' blogs for a total sample size of 172. We conducted a content analysis of these documents, and our iterative coding process focused on the framing of issues related to firm size and sustainability standards.

7.4.2 RESULTS

7.4.2.1 COMMITMENT TO FAIR TRADE VOLUME AND PERCENTAGE OF TOTAL VOLUME

Figure 4 depicts the commitment of each of the three firms to fair trade purchases during the period 2004 to 2011. It shows both total green coffee purchases and fair trade purchases for each firm, thereby illustrating the percentage of fair trade volume relative to total volume in recent years. Figure 4a shows that Starbucks has fluctuated in its commitment to fair trade: after a gradual initial increase, its fair trade purchases leveled off from 2006 to 2008, then approximately doubled in 2009 to reach 10.6% of sales, but have fallen to approximately 8% in the last two years. Green Mountain Coffee Roasters, in contrast, has steadily increased its volume of Fair trade purchases, as seen in Figure 4b, although in the last two years the percentage of fair trade purchases has declined as its total coffee purchases grew dramatically. Since 2010, GMCR has been the world's largest purchaser of fair trade certified coffee, despite purchasing less than half as much total coffee as Starbucks. Figure 4b indicates Equal Exchange's fair trade purchases were larger than GMCR in 2004 and 2005, although currently its total fair trade purchases are much less than the other two firms. Equal Exchange has maintained its commitment to 100% fair trade over this period, and its coffee purchases increased steadily from 2004 to 2009, although they have leveled off since then.

7.4.2.2 COMMITMENT TO FAIR TRADE STANDARDS

Each firm had a different response to Fair Trade USA's exit from the international fair trade system, and to its move to allow coffee from plantations to be certified as fair trade. These changes alter the social and economic aspects of sustainability that are captured by the fair trade label, which are given more prominence in both the standards and discourse of fair trade than its environmental aspects. FTUSA's actions required firms in the U.S. coffee industry to decide how to respond, and also generated a very public dialogue on the normative issues embedded in the firms' strategies.

Equal Exchange had the most dramatic reaction, announcing that it was abandoning FTUSA certification completely (the firm had already moved to alternative certifications such as IMO's Fair for Life for some of its products). Equal Exchange is currently purchasing coffee beans certified by Fair for Life, Fair Trade International (which is moving into the U.S. to compete with its former licensee FTUSA), and the Mexican nonprofit FUNDEPPO (which recently developed a Small Producers' Symbol).

Green Mountain Coffee Roasters has remained with FTUSA, and is currently sourcing coffee from one of their plantation pilot projects. Notably, the company has said it will not brand plantation coffee as fair trade "until we have evaluated the impact of these pilots at origin" [57]. Company representatives have emphasized that their measures of success include the potential impacts on sustainability, such as continuing to support small farmers. However, Larry Blanford, GMCR's president and CEO, signed a statement supporting the changes at Fair Trade USA.

Starbucks has also remained with FTUSA, but has not stated whether it will label plantation coffee as "fair trade". Ben Packard, the company's vice president for Global Responsibility, emphasized the market impacts, not sustainability goals, as a key driver of future decisions when he said, "We are waiting to see what the implications of these different schemes on the marketplace are going to mean" [11]. By "different schemes" he was also referring to Starbucks' first party C.A.F.E. Practices program, which, as mentioned above, is prioritized over fair trade in its marketing.

Equal Exchange eventually went even further than its initial response by encouraging GMCR and Starbucks to leave FTUSA certification and move to Fair Trade International's more rigorous seal. After meeting privately with GMCR executives in March, 2012 and failing to convince them of the strategy, Equal Exchange took the unusual move of placing a full-page color advertisement in Vermont's largest daily newspaper to repeat this request publicly [13]. The advertisement claimed that FTUSA's actions "threaten to reverse decades of hard-won gains while potentially putting at risk the very survival of the farmer cooperatives" [59]. Equal Exchange's Rodney North underscored the importance of this point, noting that "small farmer co-operatives were the co-creators of fair trade; to now take it and change it in a way that threatens their viability is unacceptable" [60]. Equal Exchange representatives described FTUSA's actions with respect to fair trade standards with words including "weaken," "loosen" "lower," "dilute," "undermine," and even "betrayal." While most of the firm's discourse focused on the impact the changes would have on small farmer cooperatives, it also mentioned the threat to consumer confidence in the FTUSA label. These points included that the change would make it impossible to distinguish between plantation and producer cooperative coffee, which ultimately may cause consumers to feel misled.

GMCR and Starbucks have been much less visible in the media coverage surrounding Fair Trade USA's recent moves, but it is difficult to disentangle the perspectives of the certifier and these two corporations. In explaining the standards changes, Fair Trade USA's president and CEO, Paul Rice, has frequently referred to the needs of these and other large firms, such as Dunkin' Donuts and Whole Foods (owner of the Allegro coffee brand). He responded to suggestions that roasters at these corporations replace plantation coffee with smallholder-grown coffee in their blends as "the height of arrogance" because "no one tells them how to put together the perfect French roast [10]." Dean Cycon, owner of the small, mission-driven coffee firm Dean's Beans, said that Starbucks and Green Mountain would be able to become 100% fair trade certified under the new standards without changing "their business practices one iota [14]". Even Fair Trade USA's farmer cooperative assistance program (which was announced along with its decision to certify plantation coffee) is designed to help those cooperatives that are already some of the largest, and which supply GMCR and Starbucks [10].

Two arguments frequently used by both Fair Trade USA and GMCR representatives in support of the standards changes are: (1) they will increase the positive impacts of fair trade through higher volumes, and (2) expanding or extending the scheme to hired laborers who were previously "excluded" from fair trade will help the "poorest of the poor." GMCR coffee buyer Ed Canty, for example, asked, "Why can't these workers, who are some of the poorest of the poor, in some of these [coffee] estates be involved as well [12]?" Phyllis Robinson, Equal Exchange's education and campaign manager, responded to the first point, saying that, "To the extent that more volume equals more income (and more premiums), this is good for farmers. To the extent that more volume increases the visibility of fair trade for consumers, this is also a good thing…This issue however, isn't one of volume alone…we're talking about long-term structural change [61]." Equal Exchange's blog included a statement from Marike de Pena, vice president of the Latin American and Caribbean Network of Small Producer Organizations, who challenged the second point by claiming that "farmers' incomes are far below workers' income [62]."

GMCR has increased the visibility of fair trade, however, through a "Better Quality for All" marketing campaign, as well as a more recent, celebrity-driven campaign called "Great Coffee, Good Vibes, Pass It On." These campaigns emphasize how easy it is for consumers to participate in fair trade through a "simple purchasing decision" or a "simple choice." Equal Exchange representatives, in contrast, emphasize that consumer engagement in fair trade is far from simple, suggesting it is "time consuming" and "resource heavy," but that consumers need to "go further and learn more" and "get smarter and smarter."

Equal Exchange acknowledges not only the challenges for consumers, but the difficulty that increasing scale presents for their firm and for the industry. Marike de Pena, for example, who has experience competing with plantations in fair trade as a small-scale banana farmer, wrote on the Equal Exchange blog that "it is a lot easier, cheaper, faster and safer to buy from the big estates" [62]. Robinson echoes this point in several other Equal Exchange blog posts, stating that "working with small farmer organizations can be challenging and time consuming" and that "it may be a slower and less glamorous path to work with small farmer co-ops and consumers," but this approach is likely have a greater long-term impact on sustainability [61,63].

7.5 CONCLUSIONS

Equal Exchange, Green Mountain Coffee Roasters and Starbucks were identified as the three firms accounting for the highest proportion of fair trade coffee purchase in the U.S. These firms demonstrate an inverse relationship between firm size and commitment to sustainability ideals. Equal Exchange, the smallest firm, has maintained its historical commitment to 100% fair trade coffee purchases as it has grown to be the largest of the early entrants. The firm is also attempting to lead an exodus of roasters from Fair Trade USA, which it views as weakening the previous standards for coffee by allowing certification of plantations. Starbucks, the largest firm, has increased its volume commitment to fair trade over time, but with the exception of 2009, this amount has not surpassed 8% of its total coffee purchases In addition, it has remained in the FTUSA system and has devoted few resources to raising the visibility of fair trade. GMCR is smaller than Starbucks, but recently replaced it as the world's largest purchaser of fair trade certified coffee, and this volume has approached one third of GMCR's total coffee purchases in some years. GMCR has, in contrast to Starbucks, devoted significant resources to promoting fair trade to consumers. On the other hand, the firm has also remained with FTUSA, and has offered only a mild critique of the certifier's recent policy shift by taking a "wait and see" approach on the inclusion of plantations.

While the greening of larger firms can potentially make an impact on industry sustainability more quickly than the growth of smaller, mission-driven firms, there are significant risks with this approach. In this study the larger, profit-driven firms demonstrated lower levels of commitment, a greater power to shape industry norms to their benefit, and less willingness to acknowledge tensions between size and sustainability. Although some studies suggest that large and small firms both have an important role to play in the transformation of industry toward sustainability [4,6], such a model underemphasizes the potential for co-optation [20], or for weakening standards so significantly that they eventually reduce consumer confidence in sustainability marketing as a whole [18].

These findings add to our understanding of sustainability in the fair trade coffee industry in the U.S., although their applicability to other sectors beyond coffee is somewhat limited by the nature of the case study; therefore we have primarily suggested hypotheses to be tested in other industries. The results indicate, however, that efforts to increase sustainability marketing for other products and services should take a more critical approach to pathways toward increasing industry sustainability that rely primarily on the participation of large, profit-driven firms. Attracting more smaller, mission-driven firms may be a slower approach, but may also be more likely to result in a deeply transformative effect on industry-wide sustainability efforts. Future research could identify cases of food ecolabels where barriers to the entry of large, profit-driven firms have been stronger (e.g., biodynamic, Animal Welfare Approved, and Certified Naturally Grown), and the impact that these have had on industry sustainability, as well as on consumer confidence in sustainability marketing in general.

REFERENCES

1. Boström, M.; Klintman, M. Eco-Standards, Product Labelling and Green Consumerism; Palgrave Macmillan: Basingstoke, UK, 2008.
2. Howard, P.H.; Allen, P. Beyond organic and Fair Trade? An analysis of ecolabel preferences in the United States. Rural Sociol. 2010, 75, 244–269.
3. Raynolds, L.; Murray, D.; Heller, A. Regulating sustainability in the coffee sector: A comparative analysis of third-party environmental and social certification initiatives. Agric. Human Values 2007, 24, 147–163.
4. Hockerts, K.; Wüstenhagen, R. Greening Goliaths versus emerging Davids—Theorizing about the role of incumbents and new entrants in sustainable entrepreneurship. J. Bus. Venturing 2010, 25, 481–492.
5. Davies, I.; Doherty, B.; Knox, S. The rise and stall of a Fair Trade pioneer: The cafédirect story. J. Bus. Ethics 2010, 92, 127–147.
6. Schaltegger, S.; Wagner, M. Sustainable entrepreneurship and sustainability innovation: Categories and interactions. Bus. Strat. Environ. 2011, 20, 222–237.
7. Fridell, M.; Hudson, I.; Hudson, M. With friends like these: The corporate response to Fair Trade coffee. Rev. Radical Pol. Econ. 2008, 40, 8–34.
8. Howard, P.H. Consolidation in the North American organic food processing sector, 1997 to 2007. Int. J. Sociol. Agric. Food 2009, 16, 13–30.

9. Jaffee, D. Weak coffee: Certification and co-optation in the Fair Trade movement. Soc. Probl. 2012, 59, 94–116.
10. Sherman, S. The brawl over Fair Trade coffee. The Nation, 22 August 2012. Available online: http://www.thenation.com/article/169515/brawl-over-fair-trade-coffee (accessed on 5 October 2012).
11. Gram, D. Fair Trade purists cry foul at including big farms. The Associated Press, 31 May 2012.
12. Gram, D. Too much caffeine? Fair Trade coffees fighting. The Associated Press, 21 May 2012.
13. Chesto, J. Mass market; All is not mellow with Fair Trade coffee companies. The Patriot Ledger, 23 June 2012.
14. Neuman, W. A question of fairness. The New York Times, 24 November 2011, B1.
15. Doherty, B.; Davies, I.A.; Tranchell, S. Where now for fair trade? Bus. Hist. 2012. in press.
16. Reed, D. What do corporations have to do with Fair Trade? Positive and normative analysis from a value chain perspective. J. Bus. Ethics 2009, 86, 3–26.
17. Lyons, K. Supermarkets as organic retailers: Impacts for the Australian organic sector. In Supermarkets and Agri-Food Supply Chains: Transformations in the Production and Consumption of Foods; Burch, D., Lawrence, G., Eds.; Edward Elgar Publishing: Cheltenham, UK, 2007; pp. 154–172.
18. Gillespie, E. Stemming the tide of "greenwash". Consum. Policy Rev. 2008, 18, 79–83.
19. Jaffee, D. Brewing Justice: Fair Trade Coffee, Sustainability, and Survival; University of California Press: Berkeley, CA, USA, 2007.
20. Jaffee, D.; Howard, P. Corporate cooptation of organic and fair trade standards. Agric. Human Values 2010, 27, 387–399.
21. Hatanaka, M.; Konefal, J.; Constance, D. A tripartite standards regime analysis of the contested development of a sustainable agriculture standard. Agric. Human Values 2012, 29, 65–78.
22. Templet, P.H. Grazing the commons: An empirical analysis of externalities, subsidies and sustainability. Ecol. Econ. 1995, 12, 141–159.
23. Dawson, M. The Consumer Trap: Big Business Marketing in American Life; University of Illinois Press: Champaign, IL, USA, 2003.
24. Pagell, M.; Wu, Z. Building a more complete theory of sustainable supply chain management using case studies of 10 exemplars. J. Supply Chain Manag. 2009, 45, 37–56.
25. Reardon, T.; Timmer, C.P.; Barrett, C.B.; Berdegué, J. The rise of supermarkets in Africa, Asia, and Latin America. Am. J. Agr. Econ. 2003, 85, 1140–1146.
26. Reardon, T.; Timmer, P.; Berdegue, J.A. The rapid rise of supermarkets in developing countries: Induced organizational, institutional, and technological change in agrifood systems. Electron. J. Agric. Dev. Econ. 2004, 1, 168–183.
27. Fridell, G. The co-operative and the corporation: Competing visions of the future of Fair Trade. J. Bus. Ethics 2009, 86, 81–95.
28. Haynes, J.; Cubbage, F.; Mercer, E.; Sills, E. The search for value and meaning in the cocoa supply chain in Costa Rica. Sustainability 2012, 4, 1466–1487.

29. Dougherty, D. Interpretive barriers to successful product innovation in large firms. Organ. Sci. 1992, 3, 179–202.
30. Dauvergne, P.; Lister, J. Big brand sustainability: Governance prospects and environmental limits. Global Environ. Change 2012, 22, 36–45.
31. Acs, Z.J.; Audretsch, D.B. Innovation, market structure, and firm siz. Rev. Econ. Stat. 1987, 69, 567–574.
32. Lepoutre, J.; Heene, A. Investigating the impact of firm size on small business social responsibility: A critical review. J. Bus. Ethics 2006, 67, 257–273.
33. Bacon, C.M.; Méndez, V.E.; Fox, J.A. Cultivating Sustainable Coffee: Persistent paradoxes. In Confronting the Coffee Crisis: Fair Trade, Sustainable Livelihoods and Ecosystems in Mexico and Central America; Bacon, C.M., Méndez, V.E., Gliessman, S.R., Goodman, D., Fox, J.A., Eds.; MIT Press: Cambridge, MA, USA, 2008; pp. 337–372.
34. Fair Trade USA 2011 Almanac; Fair Trade USA: Oakland, CA, USA, 2012; pp. 1–61. Available online: http://fairtradeusa.org/sites/default/files/Almanac%202011.pdf (accessed on 6 October 2012).
35. De Pelsmacker, P.; Driesen, L.; Rayp, G. Do consumers care about ethics? Willingness to pay for fair-trade coffee. J. Consum. Aff. 2005, 39, 363–385.
36. Arnot, C.; Boxall, P.C.; Cash, S.B. Do Ethical Consumers Care About Price? A Revealed Preference Analysis of Fair Trade Coffee Purchases. Canadian J. Agric. Econ./Rev. Canadienne d'Agroecon. 2006, 54, 555–565.
37. Didier, T.; Sirieix, L. Measuring consumer's willingness to pay for organic and Fair Trade products. Int. J. Consum. Stud. 2008, 32, 479–490.
38. Raynolds, L.T. Fair Trade: Social regulation in global food markets. J. Rural Stud. 2012, 28, 276–287.
39. Renard, M.-C.; Pérez-Grovas, V. Fair Trade coffee in Mexico: At the center of the debates. In Fair Trade: The Challenges of Transforming Globalization; Raynolds, L.T., Murray, D., Wilkinson, J., Eds.; Routledge: New York, NY, USA, 2007; pp. 138–156.
40. Besky, S. Can a plantation be fair? Paradoxes and possibilities in Fair Trade Darjeeling tea certification. Anthropol. Work Rev. 2008, 29, 1–9.
41. Neilson, J.; Pritchard, B. Fairness and ethicality in their place: the regional dynamics of fair trade and ethical sourcing agendas in the plantation districts of South India. Environ. Plann. A 2010, 42, 1833–1851.
42. Raynolds, L.T. Fair Trade flowers: Global certification, environmental sustainability, and labor standards. Rural Sociol. 2012, 77, 493–519.
43. Perfecto, I.; Vandermeer, J.; Mas, A.; Pinto, L.S. Biodiversity, yield, and shade coffee certification. Ecol. Econ. 2005, 54, 435–446.
44. Perfecto, I.; Armbrecht, I.; Philpott, S.M.; Soto-Pinto, L.; Dietsch, T.V. Shaded coffee and the stability of rainforest margins in northern Latin America. In Stability of Tropical Rainforest Margins; Tscharntke, T., Leuschner, C., Zeller, M., Guhardja, E., Bidin, A., Eds.; Environmental Science and Engineering; Springer: Berlin, Germany, 2007; pp. 225–261.
45. Méndez, V.E. Farmers' livelihoods and biodiversity conservation in a coffee landscape of El Salvador. In Confronting the Coffee Crisis: Fair Trade, Sustainable Livelihoods and Ecosystems in Mexico and Central America; Bacon, C.M., Méndez,

V.E., Gliessman, S.R., Goodman, D., Fox, J.A., Eds.; MIT Press: Cambridge, MA, USA, 2008; pp. 207–236.

46. Raynolds, L.T. Mainstreaming Fair Trade coffee: From partnership to traceability. World Dev. 2009, 37, 1083–1093.
47. Ware, C. Information Visualization: Perception for Design, 2nd ed.; Morgan Kaufmann: San Francisco, CA, USA, 2004.
48. Howard, P.H. Visualizing consolidation in the global seed industry: 1996-2008. Sustainability 2009, 1, 1266–1287.
49. Coffee Barometer; Tropical Commodity Coalition: The Hague, Netherlands, 2009; pp. 1–20.
50. Green Mountain Coffee Roasters Corporate Social Responsibility Report; Green Mountain Coffe Roasters: Waterbury, VT, USA, 2010; pp. 1–68.
51. Starbucks Global Responsibility Report. Starbucks: Seattle, WA, USA, 2010; pp. 1–15.
52. Fair Trade USA 2010 Almanac; Fair Trade USA: Oakland, CA, USA, 2010; pp. 1–63.
53. Obermiller, C.; Burke, C.; Talbott, E.; Green, G.P. "Taste great or more fulfilling": The effect of brand reputation on consumer social responsibility advertising for Fair Trade coffee. Corp. Reputation Rev. 2009, 12, 159–176.
54. Grodnik, A.; Conroy, M.E. Fair Trade Coffee in the United States: Why Companies Join the Movement. In Fair Trade: The Challenges of Transforming Globalization; Raynolds, L.T., Murray, D., Wilkinson, J., Eds.; Routledge: New York, NY, USA, 2007; pp. 83–102.
55. Davies, I. Alliances and networks: Creating success in the UK Fair Trade market. J. Bus. Ethics 2009, 86, 109–126.
56. Starbucks Global Responsibility Report; Starbucks: Seattle, WA, USA, 2011; pp. 1–22.
57. Green Mountain Coffee Roasters Corporate Social Responsibility Report; Green Mountain Coffee Roasters: Waterbury, VT, USA, 2011; pp. 1–75.
58. Equal Exchange Annual Report; Equal Exchange: Bridgewater, MA, USA, 2011; pp. 1–16.
59. An open letter to Green Mountain Coffee Roasters from Equal Exchange. 21 May 2012. Available online: http://www.equalexchange.coop/gmcr-ad.pdf (accessed on 6 October 2012).
60. Hill, C. Fair Trade USA's coffee policy comes under fire. East Bay Express, 25 April 2012. Available online: http://www.eastbayexpress.com/ebx/fair-trade-usas-coffee-policy-comes-under-fire/Content?oid=3184779 (accessed on 5 October 2012).
61. Robinson, P. Trying (but Failing) to Understand Arguments in Support of Fair Trade USA. Available online: http://smallfarmersbigchange.coop/2012/09/26/trying-but-failing-to-understand-arguments-in-support-of-fair-trade-usa/ (accessed on 10 October 2012).
62. Robinson, P. Banana Producers from the Dominican Republic Say no to Plantations! Available online: http://smallfarmersbigchange.coop/2012/06/14/4866/ (accessed on 10 October 2012).

63. Robinson, P. Join with us to support authentic Fair Trade. Available online: http://smallfarmersbigchange.coop/2012/01/07/join-with-us-to-support-authentic-fair-trade/ (accessed on 10 October 2012).

CHAPTER 8

Conventional to Ecological: Tea Plantation Soil Management in Panchagarh District of Bangladesh

JAKIA SULTANA, NOOR-E-ALAM SIDDIQUE, KAMARUZZAMAN, AND ABDUL HALIM

8.1 INTRODUCTION

The north-eastern parts of Panchagarh District in Bangladesh have come under tea (*Camellia sinensis* L.) plantation due to favorable soil and climate. Many small-scale farms and several tea estates have started production of tea. It has created a good avenue of employment for the deprived locals and created an opportunity of increased tea exports. The government of Bangladesh has been providing assistance for enhancing small-scale tea farming. Besides, breakthrough from private tea estates has been occurred in terms of production, processing and marketing of tea commercially from the year 2000. Still several setbacks are to overcome for tea farming which includes lack of capital, technical know-how and perennial water sources. Furthermore, low price is offered for tea leaves in case of small-scale growers and lower market value of made tea but high tariffs on external inputs are hindering the profitability. Another major concern is the significant soil degradation and hence tea productivity declining. The yield of tea is 1,100 kg/ha which is quite low as compared to other tea growing countries. Tea industry is one of the major sources of income. But it is facing a multitude of problems. For successful tea cultivation, setbacks related to soil health, management options, processing and marketing are required to address soon in Bangladesh (Islam, 2005; Ahsan, 2011; BTB 2009; RTRS, 2012).

TABLE 1: Production and product value of tea estates in the Panchagarh district

Name of Tea Estates	Production per year	Value of product	Job created man days/year	Comment
Conventional Farm				
Tetulia Tea Company Ltd. (TTCL)	180 ton tea/year	Tk. 63.00 million	371,452	Productivity stagnant
Korotoa Tea Farm (KTF)	135 ton tea/year	Tk. 47.25 million	309,213	Productivity declining
Organic Farm				
Kazi & Kazi Tea Estate Ltd (KKTE)	250 ton tea/year	Tk. 87.50 million (USD 1.27 million)	683,750	Potential is High

(Source: Rahman, 2009 and BBS, 2009)

The conventional approach of tea cultivation based on agro-chemical is causing soil degradation in the Panchagarh district of Bangladesh. The development of land for tea cultivation in the area has resulted in significant soil degradation (Ahsan, 2011); decline in soil organic matter (OM), loss of N and P through erosion and leaching, fixation of P, reduction of soil microorganisms, and acidification associated with nitrogenous fertilizers. The productivity has declined and expansion of the industry has threatened by poor conventional management practices. The traditional cultivation practices, such as excessive cultivation, continue cropping, removal of crop residues and excessive use of chemicals are contributing in land and environmental degradation. The excessive and unbalanced use of agrochemicals has led to increase production costs but decline in farm productivity. Thus, there is growing emphasis in the region for ecological and/or sustainable (integrated natural resource based farming) approach in tea cultivation to replace the conventional (chemical fertilizer based farming) approach (BTB, 2009; RTRS, 2012). Moreover, tea growers are using chemical fertilizers for higher production of tea but this approach is harmful for the productivity of tea farm. Now sustainability of conventional tea farm production in the region is under threat. Farm yield has declined substantially due to indiscriminate use of agro-chemicals and conventional practices. Thus, sustainable and or ecologically suitable management is highly demanded to sustain tea plantation in the region. However, the key research question of this project is whether integrated natural resource management is a viable alternative for the conventional soil management of tea plantations?

This case study will focus on soil management by integrated approach that will reduce the demand of external fertilizers, increase farm resource utilization and soil fertility restoration. Successful adoption of integrated approach through efficient resource management might have positive impacts on soil health, tea productivity and farm sustainability. Thus, farm economic viability and social impacts will sustain longer. The objectives of this case study were to promote productivity and sustainability of tea plantation soils at farm scale through an integrated natural resource management system, and to estimate fertilizers from organic sources that may reduce the dependency on chemical fertilizers for tea plantation in the Panchagarh district of Bangladesh.

8.2 MATERIALS AND METHODS

This case study is about integrated soil and plant nutrients management of tea plantation in Panchagarh District of Bangladesh. Information was collected through personal communication, interview with tea growers and publications (annual reports, biennial reports, books, reading materials etc.) of the Bangladesh Tea Board (BTB), Regional Tea Research Station (RTRS), Soil Resource Development Institute (SRDI) and Bangladesh Agriculture Research Council (BARC). Furthermore, it is prepared with the help of information from other secondary sources such as books, reading materials, publications, and articles found in the Wageningen University, the Netherlands library and from other sources, such as information are collected from internet sources. Additionally, relevant subjects, lectures on integrated natural resource management, field tours also remain helpful for understanding and preparing of this case study report. The estimation of fertilizer was calculated based on Fertilizer Recommendation Guide, 2005 of Bangladesh Agriculture Research Council.

8.3 RESULTS AND DISCUSSION

8.3.1 TEA PRODUCTION

Tea plantation was started with only 300 acres of land in Panchagarh District of Bangladesh but now it is being cultivated on over 2,750 acres of land. It has the potential to expand more areas of about 60,000 acres. Bangladesh Tea Board (BTB) has stated that over 16,000 hectares land is suitable for tea farming there. Currently, tea is being cultivated over 275 small gardens (3-5 acres), 17 medium sized gardens (>5 acres) and 18 big estates (>20 acres) involving over 500 small-scale farmers in the Panchagarh District. Nearly 7,500 skilled and unskilled workers have been working in tea gardens (RTRS, 2012, BTRI, 2010 and BTB, 2009). The production, product values of tea, job created and productivity status have shown in the Table 1.

TABLE 2: Mean soil fertility status of TTCL, KTF and KKTE tea farms

Name of Tea Farms	PH	OM (%)	meq/100g			TN (%)	µg/g		
			K	Ca	Mg		P	Zn	S
Conventional Tea Farm									
Tetulia Tea Company Limited (TTCL)	4.6	1.73	0.20	0.71	0.54	0.09	22.8	0.16	8.8
Korotoa Tea Farm (KTF)	4.2	1.59	0.14	0.62	0.32	0.08	21.4	0.14	7.2
*FRG critical soil test level	Strongly acidic	Low	Low	V. Low	V. Low	V. Low	Medium	V. Low	V. Low
Organic Tea Farm									
KKTE	5.1	2.73	0.32	0.83	0.61	0.16	25.3	0.32	13.6
BARC critical soil test level	Strongly acidic	Medium	Optimum	Medium	Low	Low	Optimum	Low	Medium

(*Source: ULSRUG, 2011 and RTRS, 2012), *Fertilizer Recommendation Guide-2005*

8.3.2 SOIL TYPE AND FERTILIZER REQUIREMENTS

Tea is a perennial evergreen shrub. Tea growing soils are usually acidic in nature. Soil acidity is further aggravated by the extended use of nitrogenous fertilizers (urea and ammonium sulphate) to obtain higher yield. The maintenance of an optimal soil pH (4.5-5.5) is important in tea cultivation (Natesan, 1999). Generally lime and dolomite is applied to soil as an amendment when the pH is <4.5. Furthermore, tea is a crop that takes up large quantities of A13+, thus it requires an adequate supply of exchangeable Al and Fe (Foy et al., 1978). High mortality and stunted growth of tea plant is caused by high pH and low content of exchangeable Al, Zn and Fe in soils. Adjustment of soil pH and addition of organic matter are the most common methods of decreasing P deficiencies in tea soils; it also improves soil fertility status (Zhang et al., 1997). Nutrient requirements for commercial tea production are high as the harvestable portions of tea contain the largest percentage of nutrients in the plant. N is the most important nutrient element for tea cultivation because it is required in large quantities. The next important nutrients are K and P respectively for tea plantation (Kamau, 2008; Ranganathan and Natesan, 1985).

8.3.3 SOIL FERTILITY STATUS OF TEA FARMS

The soil fertility statuses of three tea farms are shown in table 2. The soil fertility status of Tetulia Tea Company Ltd. (TTCL) and Korotoa Tea Farm (KTF) is poor than Kazi and Kazi Tea Estates Ltd. (KKTE) of Panchagarh district in Bangladesh. This might be due to the result of conventional management of TTCL and KTF. Most of tea growing farms are conventionally managed based on agro-chemicals. KKTE is the only organically managed farm in the region, thus the soil fertility status is comparatively better. However, the productivity of KTF is declining rapidly (Table 1). The decrease in crop yield is related to the decrease in soil quality of tea plantation of Korotoa Tea Farm (KTF). The soil fertility status of KTF farm is poor for most of the nutrient elements. This degraded soil has caused the production to be very low, even less than 500 kg/ha (RTRS, 2012). Among different soil and crop management factors, fertilization has substantial effect on the quality of tea farm soil. Thus appropriate fertilizer management for tea plantation is essential in order to increase per hectare yield of tea and to maintain soil productivity (Ahsan, 2011). It is evident from the quantitative soil test data (Table 2) that the overall soil quality is declining with increasing age of the tea plantations in the regions. There are evidences of acidification, decease in soil OM, N, K, S, available P and K contents in the tea farm soils of Panchagarh district (ULSRUG, 2011). The only organically managed farm of the area Kazi and Kazi Tea Estates Ltd. is maintaining profitability and farm productivity through integrated natural resource management and improved technical know-how adaptation. On the other hand, the land degradation is evident in most of the commercial tea farms of the regions which is being indicated by the yield decrease and rise of production cost. Thus, the restoration of natural soil productivity is very crucial. To maintain soil productivity and farm profitability in case of KTF, an ecological approach through integrated natural resource management is highly essential.

To address soil degradation for Korotoa Tea Farm (KTF), the dependency on chemical fertilizers such as N, P, K for tea plantations need to be reduced as early as possible. This might also increase the profitability of the KTF. A better understanding of how productivity and resource use of tea agroecosystems run through years in the farm must be analyzed. At present, KTF depend on conventional management options and the soil fertility status has become lower than required for profitable tea plantation. Due to conventional farming, i. e., agro-chemicals, excessive and continuous cultivation, the soil has become exhausted and mineral supply from soil has declined. Thus, tea plantation growth is being hampered and hence yield has been reduced substantially. The growth rate of tea plant is also disrupted by imbalance fertilization. The economic optimum rates of fertilizers have been exceeded as confirmed by the Regional Tea Research Station (RTRS) of Panchagarh district. Nitrogenous fertilizer is applied in higher rate in the conventional farms to obtain higher yield. The acidification and nutrient imbalance is triggered by excess N-applications (Bonheure and Willson, 1992). Moreover, the ecological sustainability of tea production is threatened as well. The conventional farms can maintain farm productivity at required level only through utilization of farm natural resources and other management options. Maximization of integrated natural resource use and a defined tea fertilization plan is required for KTF. Farm management should be strategic and tactical to improving profitability of tea farming business. The management options should be economically and ecologically sound as well so that sustainability prevails for tea plantations over the years. The integrated approach of fertilizer application might be a good cost effective low-input system. It is meant to increase the farm resource use efficiency by integrated natural resource management technique. Integrated natural resource management at farm scale could be a sustainable management tool for the conventional farms like KTF and many others of the Panchagarh district in Bangladesh.

8.3.4 INTEGRATED NATURAL RESOURCE MANAGEMENT (INRM)

Sustainable production integrates the idea of natural resources utilization to generate increased output and income by less or no depletion of the natural resource base. In this context, INRM maintains soils as storehouses of plant nutrients that are essential for plant growth. INRM's goal is to integrate the use of all natural and man-made sources of plant nutrients so that plant productivity increases in an efficient and environmentally suitable manner. This will ensure soil productivity for future generations. Nutrient conservation and uptake of nutrients from the soil is another critical component of INRM. Addition of fertilizer from various organic sources is supposed to prevent the physical loss of soil and nutrients through leaching and erosion, and maintenance of natural soil fertility (Groot, 2003). Green manuring, mulch application, cover crops, intercropping and biological nitrogen fixation might help to improve soil health. Organic manures such as animal and green manures substantially aid in improving soil structure and replenishing secondary nutrients and micronutrients (Peter et al., 2002). Sufficient and balanced application of organic and inorganic fertilizers is a component of INRM.

The following natural, organic and inorganic sources might consider establishing INRM for soil improvement and management in tea farms of Panchagarh district.

8.3.4.1 NATURAL RESOURCES:

The existing natural resource base (soil, water, rivers, wetland, irrigation, rain etc.) can supply nutrients to tea farm soils.

8.3.4.2 ORGANIC NUTRIENT SOURCES

- Crop residues (recycling of nutrients, addition of OM to soil etc.)
- Green manures: It will contribute to soil fertility improvement by the following ways:
    - Nutrient mobilization (take up nutrients and release through decomposition)
    - N-fixation
    - Saving nutrients from leaching
    - Organic material added to soil (incorporation of green manures)
- Organic matter management (Cow dung, Poultry manure, FYM etc.)
- Compost preparation (organic materials, water hyacinth etc.)
- Organic wastes management

8.3.4.3 MINERAL RESOURCES

Judicial application of inorganic fertilizer depends on soil test values and fertilizer recommendation. For example, dolomite application when soil acidification becomes a problem and it suppose to increase nutrient availability (Ca and Mg) also.

8.3.5 ORGANIC MATERIAL SOURCES AND THEIR MANAGEMENT AT FARM SCALE

The major possible sources of organic matter at farm level in Panchagarh District include animal manure (dairy), crop residues (farm other seasonal crops such as rice, wheat, maize, potato etc.), Farmyard wastes (animal dung and urine, feed/fodder refuse, harvested crop residues, poultry excreta etc.), bio slurry from farm biogas plants, green manuring practices and other organic wastes of various kinds such as water hyacinth available in the wet lands of the region (FRG, 2005).

Animal manure

It might include the excreta (dung and urine) from the farm animals and also from the animals of farm cooperative system, even collection of cow dung from individual farmer is also possible. Animal manure should be stored in pits preferably under a roof. The urine of cattle is rich in nitrogen and should be preserved with the dung. The manure in the pit should be kept moist in order to reduce the volatilization of nitrogen in gaseous forms. In the Panchagarh district, poultry industry is extensive and established; thus poultry excreta is cheaply available and in large amounts. Use of poultry manure for tea plantation could be an economic source of fertilizers. The conventional farm management could develop on farm dairy which might be good source of cow dung and FYM. It can also be developed through dairy cooperative by involving the rural people of the area. The only organic tea estates in the region, namely Kazi and Kazi Tea Estates Ltd. (KKTE) has developed a good on farm and cooperative dairy system and engaged the deprived community of the region for their economic benefits.

Crops residues

Leftover parts of crops after harvest are called crop residues. Crop residues of all kinds including roots, straw & stalks and vegetable tops are a source of organic matter and plant nutrients. Crop residues can be recycled either by composting or by mulching or by direct incorporation into the soil. Tea is not the only crop grown in the area; other seasonal crop cultivation might good source of income and crop residues such as rice, maize, potato etc.

Compost

The organic fertilizer that is produced by decomposing different waste materials of plant and animal origin is called compost. Ingredients that might be used to make compost in tea farms include cow dung, dead leaves, straw, weeds, water hyacinth, fruit and vegetable parts, sugar mill bagasse, rice husk etc. There is a good source of sugar mill bagasse in the region as few sugar mills are running in Panchagarh district.

Green manure

Green manure adds substantial quantities of organic matter and nitrogen to soils. Any herbaceous plant may be used for green manuring but plants of the family leguminosae are preferred because of the added advantage of getting fixed nitrogen. The common GM plants include Dhaincha (*Sesbania aculeata*), Sunhemp (*Crotalaria juncea*), Cowpea, Grasspea, Soybean, Mungbean, Blackgram etc. A green manure crop may add 10 - 15 ton of biomass (fresh weight) and 60-90 kg of N per hectare to the soil (FRG, 2005; RTRS, 2012).

8.3.6 ESTIMATIONS OF FERTILIZERS FROM ORGANIC SOURCES AT FARM SCALE

Tea plantation fertilizer requirement at the young and mature growth stage are shown in appendix 1 and 2. The fertilizer requirement for young and mature tea plantation that might be possible to meet up from various organic resources are estimated in the Table 3 and 4 respectively for the Korotoa Tea Farm (KTF). Thus, additional fertilizer requirement from external source can be determined for KTF.

TABLE 3: Fertilizer requirements and estimation for Young Tea (first year*) of Korotoa Tea Farm (KTF)

	N	P	K
1.Recommended dose** (Kg/ha), Appendix 1	80	40	80
2. Supply from organic sources 7 ton/ha	Options*** (Appendix 3)		
Cow dung (decomposed)	31.5	10.5	35
FYM	21	4.9	17.5
Poultry manure	80.5	73.5	49.0
Compost	17.5	7.0	21
Choice	Poultry Manure		
1. To add from chemical fertilizers (1-2)	-0.5	-33.5	+31.0

*the second year fertilizer requirement is about 10% higher. **the recommendation has been made on the basis of production (about 1000 kg/ha) of made tea yield. The fertilizer requirement is more for higher yield goals. *** The organic sources need to be defined by the farm authority depending on resources and farm situations.

TABLE 4: Fertilizer requirements and estimation for Mature Tea* (third year) of Korotoa Tea Farm (KTF)

	N	P	K
1. Recommended dose (Kg/ha) Appendix 2	50	4.5	25
2. Supply from organic sources 5 ton/ha	Options (Appendix 3)		
Cow dung (decomposed)	22.5	7.5	25
FYM	15	3.5	12.5
Poultry manure	57.5	52.5	35
Compost	12.5	5.0	25
Choice	Cow dung		
3. To add from chemical fertilizers (1-2)	+27.5	-3.0	0

Tables 3 and 4 show estimated amounts of NPK fertilizers which might be available by the efficient utilization of farm resources. For the computation of contribution from organic materials as fertilizer source the 'Fertilizer Recommendation Guide (FRG)' published by the 'Bangladesh Agriculture Resource Council (BARC)' has been consulted briefly. To meet up the fertilizer demand of tea plantation, a systematic organic manuring schedule is required right from the nursery stage up to maturity. The doses of fertilizers application might vary according to the age of plantation, type of tea, its performance, soil fertility and the yield goals of tea.

8.3.7 LANDSCAPE SCALE OF INRM APPROACH

The lands for tea plantations in Panchagarh district were totally fallow over last few decades. Tea plantations are creating green carpet on the vast and fallow sandy areas. Through integrated natural resource management, it is possible to cover the area under green cultivation for tea sustainably. The community people might use natural resources for synergizing various interests and activities in the development of partnerships and actions for social changes. The integration of INRM with other terrestrial ecosystem is important to sustain INRM approach in a landscape. Agricultural practices and farm management also have an impact on different plantation (Groot, 2012). However, a nature friendly and or ecologically viable farming activity might spread out in the region if INRM technique becomes acceptable and sustainable. Thus, community people might be benefitted either working with tea farms directly or by engaging as suppliers by supplying various inputs including organic materials (cow dung, crop residues), sand, top soil for nursery, stone for the construction works, dairy feed, straws, bamboo and other materials for farming activities. The barren landscape might set a good example of socioeconomic benefits for local, regional or national levels through the contributions in economic growth, poverty reduction, human development and green farming (Islam, 2012). Socio-economic upliftment and livelihood development is the scope of landscape level by adoption of INRM for tea plantations.

8.3.8 SOCIAL IMPACTS OF TEA FARM PROFITABILITY

Impacts that has been created by tea farming includes social (education, health, food and income security, social services, youth development), economic (community change and livelihood development, community services, employment) and environmental (greening land and environment) dimensions (Islam, 2012). The overall socio-economic condition of common people is changing fast in Panchagarh district following a faster growth of the tea sector. At the same time, hundreds of Panchagarh district's females, who lived in utter miseries due to abject poverty for years together, are now changing their fate and achieving self-reliance by earning wages as plucking workers in the dozens of tea gardens at the officially recognized third tea zone of the country. The growing tea sector in the Panchagarh district has boosted in a new hope for enhancing the standard of socio-economic life and woman empowerment.

8.4 CONCLUSION

It is evident from the estimation of fertilizers that the requirement of chemical fertilizers might be avoided or minimized by the adoption of INRM based on organic sources of fertilizers and OM management. The chemical fertilizer requirements for young tea might be avoided by the use of poultry manure; the total amount of N and P requirement might be possible to meet up from poultry manure. While the fertilizer requirements for mature tea might be avoided or minimized by the use of cow dung; almost 50% of N requirement and total amounts of P and K requirements might be possible to meet up. These may increase the resource use efficiency and farm profitability. It might be concluded that INRM has the potential to promote productivity and sustainability of soils under tea plantation, and organic sources are meant to reduce chemical fertilizer requirements substantially for tea plantation in the Panchagarh district of Bangladesh.

REFERENCES

1. Ahsan, Q. (2011). Maximizing yield of tea in some selected problematic acidic soils through improved management practices. Bangladesh Tea Research Institute (BTRI), Bulletin No. 17, 2011, Center for Agri-Research and Sustainable Environment and Entrepreneurship Development (CASEED), Dhanmondi, Dhaka.
2. BBS (2009). Statistical year book of Bangladesh. Bangladesh Bureau of Statistics.
3. Bonheure, D., Willson, K.C., 1992. Mineral nutrition and fertilizers. In: Willson, K.C., Clifford, M.N. (Eds.), Tea: Cultivation to consumption. Chapman and Hall, London, pp. 269-329.
4. BTB, (2009). Annual Report-2009. (http://www.teaboard.gov.bd/index.php?option=historyteaarea)
5. BTRI, (2010). Biennial Report. Bangladesh Tea Research Institute, Government of people's republic of Bangladesh, pp. 94-98, Srimangal, Moulvibazar, Bangladesh.
6. Foy, C. D., Chaney R. L. and White M.C. (1978). The physiology of metal toxicity in plants. Ann. Rev. Plant Physiol. 29: 511-526.
7. FRG, (2005). Fertilizer Recommendation Guide, Bangladesh Agriculture Research Council (BARC), Dhaka. (www.barc.gov.bd)
8. Groot, J. C. J. (2003). Exploring the potential for improved internal nutrient cycling in dairy farming systems, using an eco-mathematical model. NJAS, pp. 165-194.
9. Groot, J., (2012). Lecture on Organic agriculture, biodiversity & environmental services at the farm and landscape scales. Wageningen University, Netherlands.
10. Islam G. M. R. (2005). Present status and future needs of tea industry in Bangladesh. Pakistan Acad. Sci. 42(4) : 305-314.6. ITC. (2001). International tea committee report.
11. Islam M. S., (2012). Measuring Impact of Kazi & Kazi Tea Estate Limited, Panchagarh-An Organic Garden in Bangladesh. IOSR Journal of Business and Management (IOSRJBM), ISSN: 2278-487X Volume 3, Issue 3 (Sep-Oct. 2012), pp 01-09. (www.iosrjournals.org)

12. Kamau, D. M. (2008). Productivity and resource use in ageing tea plantations. PhD thesis, Wageningen University, Wageningen, The Netherlands.
13. Peter G. Francesco G. and Montague Y. (2002). Integrated nutrient management, soil Fertility and sustainable agriculture: current issues and future challenges. Food, Agriculture, and the Environment Discussion Paper 32. International Food Policy Research Institute, 2033 K Street, N.W. Washington, D.C. 20006 U.S.A.
14. Natesan, S. (1999). Tea Soils. In: Global advances in tea science (Ed N. K. Jain), pp 519-532 (Aravali Books International Pty Ltd, New Delhi).
15. Rahman, M. M. (2009). Green job assessment in agriculture and forestry sector of Bangladesh. Final report of International Labour Organization, Dhaka 1209, Bangladesh. International Labour Organization. (www.ilo.org/publns)
16. Ranganathan, V. and Natesan. S. (1985). Potassium nutrition of tea. In: Munson R.D. (ed). Potassium in Agriculture. ASA-CSSA-SSSA, USA. Pp. 981-1015.
17. RTRS, 2012. Regional Tea Research Station, Annual Report, 2011, Panchagarh. Pages – 7 to 19.
18. ULSRUG, (2011). Upazila Land and Soil Resource Utilization Guide, Tetulia, Food Security Programme, 2006, Soil Resource Development Institute, Ministry of Agricultrure, Bangladesh. (www.fssfc.gov.bd, www.srdi.gov.bd)
19. Zhang, Y. L., Luo, S. H., Zeng Y. H. and Pengo P. Y. (1997). Study nutrient scale of sufficiency or deficiency in tea soils in Hunan province and fertilizing recommendation. J. of Tea Soil 17 (2): 161-170.

There are several supplemental files that are not available in this version of the article. To view this additional information, please use the citation on the first page of this chapter.

PART V

FOOD AND BEVERAGE
WASTE PRODUCTS

CHAPTER 9

Energetic Analysis of Meat Processing Industry Waste

COSMIN MĂRCULESCU, GABRIELA IONESCU, SIMONA CIUTĂ, AND CONSTANTIN STAN

9.1 INTRODUCTION

The meat processing industry generates meat and products for human consumption, and also considerable amounts of solid waste and other by-products (skins, fats, bones, offal, etc.) [1, 2, 3]. A significant percentage from the animal remains after processing as waste: approximately 46% of each cow, 48 % of each sheep or goat, 38 % of each pig, 28 % of each chicken [4]. These by-products are a hard to use waste that is produced each year worldwide in millions of tons [5]. The most common method adopted by the meat processing industry to dispose of animal waste is the processing into a meat and bone meal [6].

The Meat and Bone meal (MBM) is produced by processing at 105 °C the animal's offal, blood and bones previously mixed and crushed [7]. Be-

Reprinted with permission from the journal. Energetic Analysis of Meat Processing Industry Waste. © *Mărculescu C, Ionescu G, Ciută S, Stan C.* UPB Scientific Bulletin, Series C *75.2 (2013),* http://www. scientificbulletin.upb.ro/rev_docs_arhiva/fulld8a_200720.pdf.

fore 1994 in the EU, the MBM was used for feed the ruminants. Because it was found to be the main cause of bovine spongiform encephalopathy (BSE) disease transmission the European legislation has banned the MBM use as feed [8]. Thereby, the safe disposal of meat processing industry waste is necessary to prevent hygienic problems and pathologically transmissible diseases. Landfill is not a viable solution because it cannot destroy any potential pathogens.

Nowadays, it is seen that the most potential alternative for disposal of this waste is thermal treatment by high-temperature processes [9]. One option can be the combustion and gasification with energy recovery potential [10]. This paper presents the experimental results in term of physical-chemical characterization of meat processing industry residues (bones) in order to assess their energetic potential as renewable energy source. Based on experimental analysis were identified possible energy conversion solutions.

9.2 MATERIALS AND METHODS

For the optimum waste processing technology configuration the base of the project consists in the complete product characterization with respect to proximate and ultimate analysis, energy content and physical-mechanical properties. The last criteria is required by collecting, transport, storage and feed-in procedures related to waste treatment/valorization unit.

The samples used for the characterization of waste (mixtures of bones and meat) were collected directly from the meat processing line. The sampling, samples preparation and laboratory experiments/analysis were conducted in order to insure the results reliability.

9.2.1 PROXIMATE ANALYSIS

Proximate analysis delivers the main physical component distribution resulted under controlled heating, in the following order: moisture, volatile matter, fixed carbon, and ash.

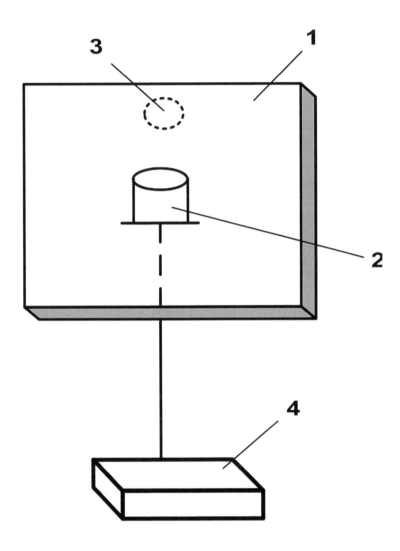

FIGURE 1: Electric heated oven 1. Electrically heated oven, 2. Steel crucible, 3. Flue gas evacuation, 4. Analytical balance

The moisture content was determined by sample drying at 105°C for 24 hours. Due to fat based components in the sample the temperature level may induce supplementary weight loss by partial volatilization. Consequently the moisture content may be slightly lower than the measured one.

For determination of volatile matter, the waste was heated to 800 °C for 40 minutes in inert atmosphere; an electric heated oven was used. In Figure 1 is presented the electrically heated oven.

The sample is introduced in a steel crucible with a cap to prevent the oxidization. The mass variation of sample can be monitored with an analytical balance. Volatile matter is the material that will convert to gases. The devolatilization was considered completed when the variation of mass sample stops. The remaining material at the end is char.

For inert (non-combustible fraction) determination, the char was heated in the oven at 1000 °C without the cap for 30 minutes until the complete combustion takes place. The crucible containing the sample is removed and cooled for 15 minutes before it is weighed, the remaining material is ash.The fixed carbon was determined by difference and represents the solid carbon in the sample that remains in the char after the complete volatile release.

9.2.2 ULTIMATE ANALYSIS

The elemental composition of the waste was determined experimentally using an elemental analyzer. The apparatus uses the principle of dynamic flash combustion and gas chromatography separation of the resultant gaseous species with thermal conductivity detection for measuring carbon, hydrogen, sulfur and nitrogen.

9.2.3 ENERGY CONTENT

The energy content of a fuel is expressed as higher heating value (HHV) and lower heating value (LHV). The HHV was determined experimentally by a calorimetric system, by burning a representative sample in the calorimetric bomb. The LHV is computed using a correction equation based on hydrogen and water content of the sample [11]:

$$LHV = (HHV - 5.83 \times W) \times 4.18 \ [kJ/kg] \tag{1}$$

Where W is the percentage of water content in sample, calculated as:

$$W = W_t + 9 \times H \ [\%] \tag{2}$$

where:

- W_t: total humidity content,
- H: mass percentage of hydrogen

Based on experimentally elementary composition and using semiempirical formulas, the calorific value of sample was also determined.

The high heating value (HHV) was calculated using Dulong formula [12]:

$$HHV_{dry \ product} = (7831 * C) + [35,932 * (H - O_2/8)] + (1187 * O_2) + (578 * N) \tag{3}$$

A second formula specially designed by Channiwala and Parikh for biomass based products was used [13].

$$HHV_{dry \ product} = 349.1 \times C + 1178.3 \times H + 100.5 \times S - 103.4 \times O - 15.1 \times N - 21.1 \times A \tag{4}$$

Where:

- C, H, O, S, N: is the percentage of these chemical elements in the composition of the sample and A is the inert fraction.

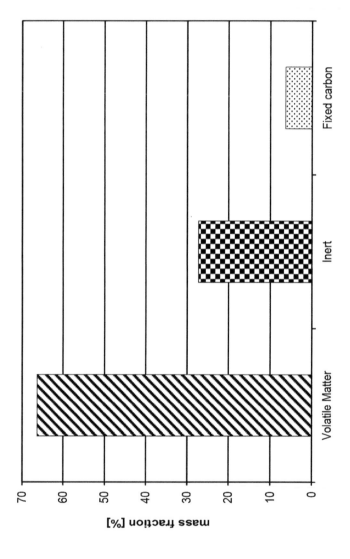

FIGURE 2: Proximate analysis of sample

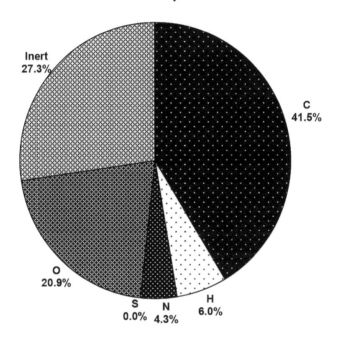

FIGURE 3: Ultimate analysis of pork bones

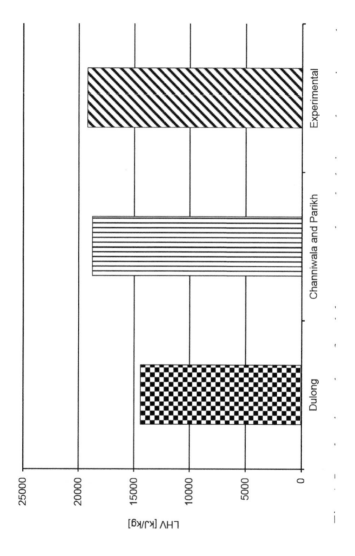

FIGURE 4: Low heating value of pork bones waste by calculation and experimental

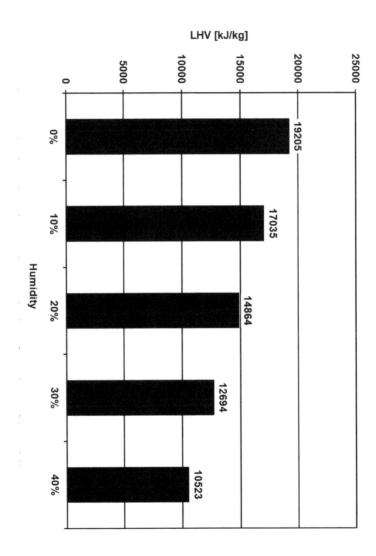

FIGURE 5: The low heating value variation function of humidity

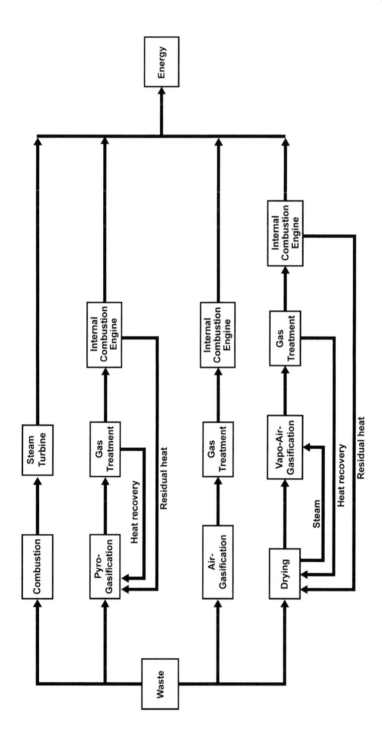

FIGURE 6: Energy chain options

• The low heating value (LHV) is calculated with formula (1) considering that the product is completely dried.

9.3 RESULTS AND DISCUSSIONS

9.3.1 PROXIMATE ANALYSIS

Figure 2 presents the values obtained for the volatile content, fixed carbon and inert of the sample completely dried. The volatile content of the sample is very high, up to 67% similar to fast biodegradable biomass products. The presence of important volatile matter guarantees a good ignition point and a lower excess oxygen demand for a complete burning process. A specific characteristic of this product is given by high inert fraction (28%) different from common biodegradable waste. This value is similar to coals and municipal solid waste (only for non-selective collection situation) [11].

The presence of high inert fraction may raise residues management problems whatever the waste to energy conversion solution will be applied. Regarding the fixed carbon presence in the product its mass fraction does not exceed 7–8%. Consequently, the specific energy content of the char is very low. The product energy content is concentrated in its volatile matter.

9.3.2 ULTIMATE ANALYSIS

In Figure 3 is presented the results of the ultimate analysis of pork bones. The combustible fraction from the sample, C – 41.5% and H – 6% are high, confirmed by the high calorific value. The sample has also a significant content of N – 4.3%. No sulphur or chlorine elements were detected. The oxygen concentration is similar to biomass based products.

In terms of pollutant sources the product is considered clean suitable for thermal-chemical processing. The waste to energy conversion technologies will require reduced gas cleaning units and lower investment costs.

9.3.3 HIGH AND LOW HEATING VALUES

Figure 4 displays the results of the low heating values calculated by the elemental composition and from experimental determination. The calculated values are presented to highlight the importance of direct measurement of HHV in LHV determination. Important errors are introduced by empirical formulas used for HHV and LHV calculation even if correct measured values for elemental composition of the product are used.

By calculation using two different formulas the low heating value varies from 14424 kJ/kg with Dulong to 18755 kJ/kg with Channiwala and Parikh. The second formula delivers more accurate results compared to experimental determination (19205 kJ/kg). Nevertheless there is no Romanian standard to certify the results from calorimeter measurements.

For thermal-chemical processing the low heating value and water content of the products represent important data that strongly influence the design and the operation parameters of the waste processing units. In this case lower water content, achievable by pre-treatment or dying, will improve not only the product LHV but also the energy efficiency of the thermal processing. This will lead to higher global efficiency of the conversion chain.

The influence of water content of the product on its LHV is presented in Figure 5. The presence of water in the product up to 40% (in mass) will decrease the low heating value down to 10523 kJ/kg. Even so the value is relatively high in range with medium quality coals. For example the lignite from Oltenia extraction region does not exceed 7000 kJ/kg [14].

Nevertheless the options for moisture removal are limited for this product due to its physical characteristics that are not suitable for mechanical grinding at mesh compatible sizes.

9.4 ENERGY CONVERSION

Based on primary, ultimate and calorimetric analysis results a series of waste to energy conversion chains were assessed. The conversion solutions are presented in Figure 6. The first option is represented by direct combustion combined with steam turbine cycle. The waste calorific value and its elemental composition are suitable for this process. Some important aspects related to product properties and in-field conditions limit the applicability of this solution:

- High inert fraction. The presence of important ash fraction and its low melting point will limit the maximum combustion temperature. Consequently the steam parameters will be low and the turbine efficiency will decrease.
- Mass flow. Usually the waste sources have small capacities and the available quantities of waste are reduced with respect to energy sector applications. The Rankine-Hirn cycles are strongly influenced by the scale effect and for this application the global efficiencies are very low.
- Availability. Generated by an industrial process, the waste will not be delivered continuously to the power plant. Steam turbines are designed for a limited start stop procedures. Moreover waste storage is impossible due to its fast degradation properties and high health risk.

The other three options use gasification in different processing configurations. Due to Otto-Diesel cycles flexibility, fast start-up and insignificant scale effect, the gasification provides clear advantages when compared with combustion for this type of fuel and capacity.

Pyro-gasification could be used to solve the moisture problem enabling also the storage of the pyrolysis by-product. Considering the very low fraction of carbon in pyrolysis char the gasification of this product may not produce the required amount of syngas to be used in an internal combustion engine.

The direct air gasification could represent a solution but the increased air/fuel ratio required by the important moisture content within partial combustion of waste could decrease the syngas low heating value to engine operating limits. Experimental research is required.

An interesting solution may integrate a partial drying of the waste with vapor-gasification processing for high quality syngas production. The separate drying will solve the moisture presence in first gasification stage providing the steam for the second stage where hydrogen formation will be favored. The advantage is given by the significant increase of syngas quality that may fit the engine requirements.

9.5 CONCLUSIONS

The paper presents the results of experimental campaign conducted on the main meat industry residue (mixture of bones, meat etc.) for product complete characterization with respect to combustible properties. The results reveal important volatile matter (up to 67% in dry basis product) and ash content (up to 28%). The fixed carbon fraction is low compared to biomass and coals and does not exceed 7%. The elemental analysis shows results similar to biomass based products: C – 41.5%; H – 6%; O – 20.9%; N – 4.3%. No sulphur or chlorine elements were detected. The low heating value of dried product is 19205 kJ/kg while the raw product LHV does not exceed 12000 kJ/kg. The results prove that meat industry residues that consist mainly of mixtures of bones and meat could be a renewable energy source with direct in-site valorization. Based on product properties and research team experience in the field of thermal-chemical processing a series of waste to energy conversion chains is proposed.

REFERENCES

1. P.T. Williams, Waste Treatment and Disposal Second Edition, John Wiley & Sons Ltd, 2005
2. Food & Agriculture Organisation of the United Nations (FAO –www.fao.org), 2011
3. EUROSTAT yearbook 2011, www.ec.europa.eu/eurostat
4. Ministry of the Attorney General, Ontario, Canada A http://www.attorneygeneral. jus.gov.on.ca/english/about/pubs/meatinspectionreport
5. I. Gulyurtlu, D. Boavida, P. Abelha, M.H. Lopes, I. Cabrita, „Co-combustion of coal and meat and bone meal", Fuel, Vol. 84, 2005, pp. 2137–2148
6. E. Cascarosa, G. Gea, , J. Arauzo, "Thermochemical processing of meat and bone meal: A review", Renewable and Sustainable Energy Reviews, Vol. 16, No. 1, 2012, pp. 942–957
7. J. Weiss, M. Gibis, V. Schuh, H. Salminen, "Advances in ingredient and processing systems for meat and meat products", Meat Science, Vol. 86, No. 1, 2010, pp. 196–213
8. European commission decision 94/381/EC of June 1994 concerning certain protection measures with regard to bovine spongiform encephalopathy and the feeding of mammalian derived protein. Official Journal of European Union 1994; L172:23–4
9. M. Ayllon, M. Aznar, J.L. Sanchez, G. Gea, J. Arauzo, „Influence of temperature and heating rate on the fixed bed pyrolysis of meat and bone meal", Chemical Engineering Journal, vol. 121, 2006, pp. 85–96

10. E. Cascarosa, L. Gasco, Meat and bone meal and coal co-gasification: Environmental advantages, Resources, Conservation and Recycling, Vol. 59,2012, pp. 32–37
11. T. Apostol, C. Marculescu, Managementul deseurilor solide (Management of solid waste) Editor AGIR, Bucuresti, 2006
12. A. Adekiigbe, Determination of Heating Value of Five Economic Trees Residue as a Fuel for Biomass Heating System, Nature and Science, vol. 10, 2012, pp. 26-29
13. J. Parikh, S.A. Channiwala, G.K. Ghosal, "A correlation for calculating HHV from proximate analysis of solid fuels", Fuel, vol. 84, 2004, pp. 487-494
14. F. Alexe, V. Angheluță, V. Athanasovici, Manualul inginerului termotehnician, Editura tehnică, Bucureti, 1986

CHAPTER 10

Laboratory-Scale Anaerobic Sequencing Batch Reactor for Treatment of Stillage from Fruit Distillation

ELENA CRISTINA RADA, MARCO RAGAZZI, AND VINCENZO TORRETTA

10.1 INTRODUCTION

Stillage is a term used for some residues from distillery plants. These liquid residues are a big problem in the sector because of the high concentration of pollutants. However, distillery plants generate not only liquid discharges but also solid residues, the treatment of which is generally based on aerobic processes (Rada et al., 2009); in the case of liquid residues, both aerobic and anaerobic treatments are adopted (Deepak and Adholeya, 2007).

The anaerobic process is widely used for treating wet residues to produce methane or bio-hydrogen (Moletta, 2005; Bouallagui et al., 2005;

Reproduced from Laboratory-Scale Anaerobic Sequencing Batch Reactor for Treatment of Stillage from Fruit Distillation. Rada EC, Ragazzi M, and Torretta V. Water Scienceand Technology 67,5 (2013). doi: 10.2166/wst.2013.611. With permission from the authors and IWA Publishing.

Wang et al., 2011; Luo et al., 2010a and b; Nasr et al., 2011; Kaparaju et al., 2010; Alkan-Ozkaynak and Karthikeyan 2011). The aim of this process is to convert a large part of the COD into biogas, thanks to the high efficiency of BOD removal (Wolmarans and de Villiers, 2002; Vlissidis and Zouboulis, 1993). The anaerobic process has been studied over the years, from many points of view, and as a result, today, both the conventional (Fannin et al., 1982; Bories et 1998) and unconventional aspects (Rada et al., 2008; Braguglia et al., 2012; Tokumoto and Tanaka, 2012) of this process are adequately known.

The work presented was developed on a lab scale as a first simplified step of an overall research aimed to set the design parameters for implementing the anaerobic digestion of stillage on a real scale. The aim of the present research was to measure the specific gas production (SGP) and gas production rate (GPR) from the anaerobic digestion of stillage under different conditions. The ASBR (Anaerobic Sequencing Batch Reactor) technology was used to develop the tests, as this process has demonstrated high stability when applied to stillage treatment (Farina et al., 2004). To this end, this paper contributes to the study of ASBR viability in the stillage treatment sector, which is often characterized by different kinds of anaerobic reactors (Melamane et al., 2007).

The properties of stillage can vary depending of the type of distillery substrate, local crop conditions, the distillery operation itself, etc. Thus, in general, results are greatly affected by particular conditions. The results obtained in this paper refer to stillage from grape distillation.

10.2 MATERIALS AND METHODS

The experimental apparatus (Figure 1) for batch anaerobic digestion tests included 3 reactors, each with a useful volume of 5 liters, equipped with:

- **Digester**: a cylindrical steel container, that is resistant to corrosion, fitted with an external jacket. Two connection pipes were joined, and led to a thermostatic bath. Two lateral openings served to draw off sludge for sampling and for total emptying. Three openings on the cover allowed the mixer rod to be inserted, the biogas to be conveyed to the meter, and the substrate to be introduced.

- **Biogas meter:** a glass beaker with a double chamber, filled with 500 ml of water, in order to separate physically the interior and the exterior of the reactor. At the beginning of a test, under equilibrium conditions, the levels are at "0", as indicated in Figure 1. Once configuration "1" (maximum liquid level) is reached, the flow from the digester is interrupted and the connection valve to the outside is opened. Re-balancing the pressure brings the water column down to the position indicted by the number "2", the vent valve is closed and the connection to the reactor is reopened. The difference in level between "1" and "2" corresponds to a preset volume and is measured each time by an acquisition program. When the connection is reopened, the water column rises by about 1 cm, re-balancing the pressure of the meter with the pressure in the digester at the time of the interruption. This situation corresponds with the limit indicated by the number "3". The volume discharged from time to time is set by calibrating the distance between the control electrodes, marked with the letters A (maximum level) and B (minimum level). A third electrode, marked C, serves as a common reference for the control processor, and therefore, must always be immersed in the liquid. This approach kept the cost of the apparatus low. It would have been possible to prepare a NaOH solution in the gas meter to monitor the CH4 production in the test. However, this option was not adopted for the presented runs, because a detailed methane generation analysis was planned for the second step of the overall research (on pilot scale).
- **A mixer:** the turns depend on the viscosity of the mixed sludge and on the friction that the mixer rods exert against the bronzes and the oil seals, which must be close-fitting to the rods in order to prevent any loss of biogas. The number of revs per minute is altered by electronically controlling the voltage output. Tests were carried out, with the sludge mixing speed kept constant.
- **Electrovalve:** it connects the digester with the biogas meter. The flow is interrupted when the liquid in the meter reaches the maximum level, when the known volume is sent outside; then it re-establishes the connection between the meter and the digester, by closing off the connection with the environment.

Each reactor was connected to an acquisition box, which allowed the process parameters to be continuously recorded and sent to a computer. The system was equipped with a thermostatic bath that kept the temperature inside the reactors constant ($35\pm1°C$). The heating element was switched on automatically, whenever the temperature of the water went below a set value; it was switched off when it went above this limit. Finally, a pump

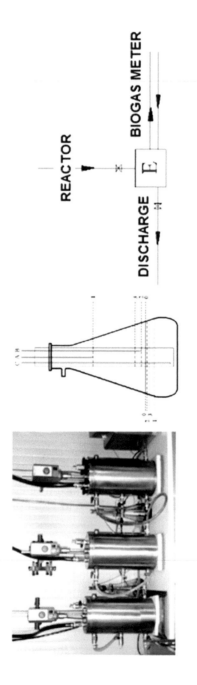

FIGURE 1: Anaerobic digestion reactors, biogas meter and electrovalve (E)

kept the water recirculating, in order to maintain the temperature of the external jacket of the digesters steady.

The main characteristics of the simplified experimental apparatus presented is that the organic substance in a single reactor can be removed, which thereby avoids the need for a separate settler and for external recirculation of the biomass. The system entails a sequence of four phases: feeding, digestion, sedimentation and discharge.

In order to be able to check the development of the anaerobic digestion of stillage, the following parameters were determined:

- Total Solids (TS): with reference to the organic part and the inert part of the substrate;
- Volatile Solids (VS): with reference to the organic part of the dry substance;
- Chemical Oxygen Demand (COD), through a $K_2Cr_2O_7$ solution in an acid environment;
- Volumetric Organic Loading Rate (OLR);
- Organic Loading Rate, with reference to the VS in the reactor (F/M);
- Gas Production Rate (GPR)
- Specific Gas Production (SGP).

$$OLR = \frac{VS_{in}}{V_{reactor}} \qquad OLR = \frac{COD_{in}}{V_{reactor}}$$

or   (1)

$$\frac{F}{M} = \frac{V_{stillage\,in}}{[VS_{reactor}]} \cdot \frac{[COD_{stillage}]}{V_{reactor}}$$

(2)

$$GPR = \frac{Q_{biogas}}{V_{reactor}}$$

(3)

$$SGP = \frac{V_{biogas}}{VS_{in}} \qquad SGP = \frac{V_{biogas}}{COD_{in}}$$

or   (4)

where:

- VS_{in}: total VS fed in (per cycle);
- COD_{in}: total COD fed in (per cycle);

- $V_{reactor}$: usable volume in the digester;
- $V_{stillagein}$: volume of stillage introduced into the reactor;
- $[COD_{stillage}]$: COD concentration of the stillages introduced;
- $[VS_{reactor}]$: concentration of the VS of the sludge present in the reactor;
- Q_{biogas}: biogas flow rate;
- V_{biogas}: volume of biogas produced in the time considered.

In the case of eq. (2), not all the VS concentration present in the inoculum (sludge) is directly microbial, so the microbial VS could be provided for a deeper analysis of the process. Another potential variation may be to due to the fact that both variables have the same unit of measure. Additional parameters measured were: ammonia and organic nitrogen, and filtered COD.

The primary aim of the tests was to measure the SGP and GPR. Three batch digestion tests were carried out with different organic volumetric loads, with a single feed and with different initial substrate concentrations; once the organic substance fed in was used up (reaching of the endogenous phase), the digesters were reloaded with an increasing target concentration of the organic substance (for several process cycles). One of the organic load values was applied for only one cycle:

- reactor R1: load cycles of stillages with increasing concentrations: 1, 3, 5, 7, 9 gCOD L^{-1};
- reactor R2: load cycles of stillages with increasing concentrations: 2, 4, 6, 8, 10 gCOD L^{-1};
- reactor R3: load cycles of stillages with increasing concentrations: 10, 15, 20 gCOD L^{-1}.

These COD concentrations (COD L^{-1}) were assessed on the basis of the initial results obtained from the analysis of the stillages before a period of storage at low temperatures (2-5°C). These values were corrected on the basis of the actual COD concentration in the stillages effectively loaded for each, individual test, obtained using specific laboratory analyses. As the stillages before the test were kept at a temperature of 2-5°C, a number of preliminary operations were carried out. The stillages were firstly warmed to a temperature of about 20°C and treated, in order to obtain a homogenous sample suitable for feeding into the reactor. In this way, thermal shock of the mesophilic biomass was avoided.

In order to carry out the anaerobic digestion tests, sludge was taken from a real anaerobic digester plant, which performed the anaerobic digestion of the excess activated sludge produced in the same wastewater treatment plant. This was used as inoculum for reactors R1, R2 and R3. This sludge was taken from time to time, and as it was not entirely stabilized, it was decided to stabilize it for 3-5 days. During this time, the sludge produced a non-negligible quantity of biogas, which could have interfered with the measurement of the biogas actually produced alone by the digestion of the stillages introduced into the reactors. Once the endogenous phase was reached, it was possible to proceed with the anaerobic digestion tests of the stillages, which were diluted according to the strategy described below. The stillages came from a grape distillery located in the North of Italy.

The subsequent tests were carried out, and the process parameters were corrected in accordance with the initial hypotheses:

- reactor R1: load cycles of stillages with increasing concentrations: 0.7, 2.6, 4.9, 7.0, 8.8 gCOD L^{-1};
- reactor R2: load cycles of stillages with increasing concentrations: 1.5, 3.4, 5.9, 8.0, 9.8 gCOD L^{-1};
- reactor R3: load cycles of stillages with increasing concentrations: 8.0, 10.7, 15.7 gCOD L^{-1}.

The monitoring program recorded the biogas production as a function of time, and determined the accumulated biogas production curve and the production rate.

10.3 RESULTS AND DISCUSSION

Table 1 shows the results for the total COD (CODtot) and the filtered COD (CODf) for the stillage and sludge used. The low CODf/CODtot ratio may be explained as a consequence of organic matter entrapment. The values for ammonia, organic nitrogen, moisture and volatile solids are also reported.

The characteristics of each individual test (one cycle each) are shown in Table 2, which highlights the fact that the assessed COD concentrations did not correspond exactly with the measured values.

TABLE 1. Characteristics of the anaerobic stillages and sludge used

			Stillages			
Code	CODtot	CODf	Ammonia	Organic N	Moisture	VS
	g L^{-1}	g L^{-1}	mg L^{-1}	mg L^{-1}	%	%
B01	85.3	48.1	139	733	97.2	93.5
B02	98.8	52.7	140	841	-	93.9
B03	114.0	52.7	176	802	-	-
B04	98.0	54.5	135	776	91.7	94.6
B05	90.1	54.9	143	747	91.8	94.7
B06	86.8	52.1	160	673	90.7	94.6
B07	86.3	51.2	169	774	92.1	94.5
			Sludge			
Code	CODtot	CODf	Ammonia	Organic N	TS	VS
	g L^{-1}	g L^{-1}	mg L^{-1}	mg L^{-1}	kg m^{-3}	kg m^{-3}
F01	11.4	2.1	713	457	24.0	19.1
F02	13.3	2.4	732	495	13.0	8.0
F03	12.4	1.9	665	492	13.5	9.3
F04	11.7	2.2	668	626	13.5	8.5
F05	11.3	3.8	626	509	11.4	7.8
F06	14.2	3.3	686	423	13.0	8.8
F07	14.1	3.3	673	297	13.3	8.7
F08	14.1	3.5	657	170	13.9	9.3
F09	13.1	3.1	644	30	14.0	9.5
F10	13.8	3.5	677	24	13.6	9.2
F11	15.4	4.2	687	102	11.9	7.5
F12	14.3	4.9	688	60	13.1	9.1
F13	16.4	4.7	647	60	14.7	10.2

All of the tests carried out followed a similar course: a quick initial production of biogas, because of the presence of rapidly biodegradable matter in the stillages; a following curve with a decreasing slope, corresponding to an ever decreasing production rate. The only exception to this course was the curve for the anaerobic digestion of stillages with an initial concentration of 15.7 gCOD L^{-1}, in digester R3: after the initial degradation phase of the rapidly biodegradable matter, a slowing of the process was detected, followed by a slow recovery and a unstable phase. The 15.7 gCOD L^{-1} proved to be unfavorable to the biological anaerobic digestion reactions of the stillages, because of factors that inhibited this process. This may depend on an excessive concentration of toxic compounds, as a consequence of the limited dilution. Indeed, stillages contain phenolic compounds which have a high antibacterial activity (Melamane et al., 2006).

TABLE 2: Typical parameters of the sludge and stillages used

Notes	stillage code	sludge code	reactor V	F / M load	assessed COD load	measured COD load
			L	$gCOD_{stillage} / gVS_{sludge}$	gCOD L-1	gCOD L-1
load 1 - R1	B01	F01	3.480	0.04	1	0.7
load 2 - R1	B02	F03	3.517	0.3	3	2.6
load 3 - R1	B03	F05	3.630	0.6	5	4.9
load 4 - R1	B04	F07	3.890	0.8	7	7.0
load 5 - R1	B04	F09	4.252	0.9	9	8.8
load 1 - R2	B01	F02	3.510	0.2	2	1.5
load 2 - R2	B02	F04	3.588	0.4	4	3.4
load 3 - R2	B03	F06	3.740	0.7	6	5.9
load 4 - R2	B04	F08	4.050	0.9	8	8.0
load 5 - R2	B04	F10	4.482	1.1	10	9.8
load 1 - R3	B05	F11	3.272	1.1	10	8.0
load 2 - R3	B06	F12	3.710	1.2	15	10.7
load 3 - R3	B07	F13	4.510	2.1	20	15.7

Figure 2 left shows good alignment of the data, giving a SGP of 281 mlgas g^{-1} COD$_{add}$ (R^2 = 0.99). The differences are emphasized by analyzing the times necessary for the digestion of the substrate fed in. The GPR was calculated considering the time necessary for total consumption of the COD

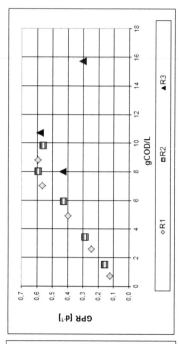

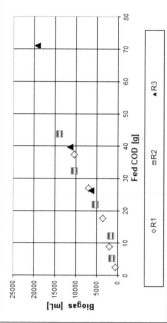

FIGURE 2: Biogas production *vs* COD (left) and GPR values *vs* initial concentration of COD (right).

introduced into the digester. There was an increase in the mean rate, up to values of 8-9 gCOD L^{-1} introduced, followed by a slight fall (Figure 2 right).

The GPR value of the first load into digester R3, was 8 gCOD L^{-1} and produced an anomalous result, possibly due to the fact that the first load suffered from incomplete acclimatization of the bacteria present in the sludge for the anaerobic digestion of the stillages.

The optimal loads for obtaining the maximum SGP and GPR values were 8-9 gCOD L^{-1} and 0.9 gCOD g^{-1}VS (load 5 R1 and load 4 R2). Under these conditions, the complete consumption of the stillages was achieved in about 4 days, as resulted from the dynamics of biogas production. By increasing the concentration of COD introduced, with similar levels of SGP and GPR, the consumption dynamics lasted longer. From these results, it is clear that dilution is a key factor for the optimization of the treatment of stillages.

Table 3 shows the accumulated production of biogas, and the biogas production rate.

TABLE 3: Table summarizing the SGP and GPR values for each test

notes	test duration	biogas produced	SGP	GPR
	days	ml	mL$_{gas}$ / gCOD	L$_{gas}$ / (V$_{reactor}$ x d)
load 1 - R1	1.142	490	191	0.123
load 2 - R1	2.283	1960	218	0.244
load 3 - R1	2.495	3621	205	0.400
load 4 - R1	3.205	7050	261	0.565
load 5 - R1	4.106	10416	279	0.597
load 1 - R2	2.638	1462	286	0.158
load 2 - R2	3.159	3248	267	0.287
load 3 - R2	3.356	5380	246	0.429
load 4 - R2	4.343	10454	323	0.594
load 5 - R2	5.575	14048	321	0.562
load 1 - R3	4.486	6365	242	0.434
load 2 - R3	5.278	11404	288	0.582
load 3 - R3	13.972	19298	272	0.306

From an analysis of these parameters (particularly from the biogas generation slopes, which show a clear peak in the GPR curve for each run), it was found that about 30% of the COD introduced was consumed within 10-16 hours after the beginning of the tests (Table 4). For these analyses, the tests with a low incoming initial COD load were discarded. Low values of COD fed in did not enhance the change in the gradient of the biogas production curve as a function of time; this made identification of this variable difficult. These tests also correspond with the initial tests, which might present problems of acclimatization of the biomass, and therefore, underestimate the measurement of the biogas production, as seen in the load 1 test in digester R3. The load 3 test in digester R3 was discarded because it presented an entirely different course from that of the other tests.

TABLE 4: Values of rapidly biodegradable COD, and removal times

notes	rapidly biodegradable COD	removal time
	% on COD tot	hours
load 3 - R1	31.4%	9.7
load 4 - R1	30.0%	12.6
load 5 - R1	34.6%	16.5
load 3 - R2	29.3%	10.2
load 4 - R2	30.0%	13.0
load 5 - R2	29.2%	11.6
load 2 - R3	30.9%	15.4

10.4 CONCLUSIONS

The work presented here demonstrates that some design parameters for an ASBR plant intended to treat stillage from fruit distillation can be established by developing a simplified lab-scale study. With a limited number of experimental runs, the optimal loads for obtaining the maximum SGP and GPR values were found: 8-9 gCOD L^{-1} and 0.9 gCOD g^{-1}VS. Under these conditions the complete consumption of the stillages can be achieved in about 4 days, as demonstrated by the biogas generation curves. Higher concentrations of COD in the feed needed longer times for completion of the process (levels of SGP and GPR were similar). Dilution was a key factor in the optimization of the stillage treatment. More specifically, around 30 of the COD was rapidly biodegradable.

REFERENCES

1. Alkan-Ozkaynak, A., Karthikeyan, K.G. 2011. Anaerobic digestion of thin stillage for energy recovery and water reuse in corn-ethanol plants. Bioresource Technology, 102(21), 9891-9896.
2. Bories, A., Raynal, J., Bazile, F. 1998. Anaerobic digestion of high-strength distillery wastewater (cane molasses stillage) in a fixed-film reactor, Biological Wastes 23(4), 251-267
3. Bouallagui, H., Touhami, Y., Ben Cheikh. R., Hamdi, M. 2005. Bioreactor performance in anaerobic digestion of fruit and vegetable wastes. Process Biochemistry, 40(3-4), 989-999.
4. Braguglia, C.M., Gianico, A., Mininni, G. 2012. Comparison between ozone and ultrasound disintegration on sludge anaerobic digestion. Journal of Environmental Management 95, S139-S143.
5. Deepak, P., Adholeya, A., 2007 Biological approaches for treatment of distillery wastewater: A review. Bioresource Technology 98(12), 2321-2334.
6. Fannin, K.F., Conrad, J.R., Srivastava, V.J. 1982. Anaerobic processes. Journal of the Water Pollution Control Federation, 54 (6) , 612-62.3
7. Farina, R., Cellamare, C.M., Stante, L., Giordano, A. 2004. Pilot scale anaerobic sequencing batch reactor for distillery wastewater treatment. X World Congress on Anaerobic Digestion, Montreal, Canada, 20/8 - 2/9
8. Kaparaju, P., Serrano, M., Angelidaki, I. 2010. Optimization of biogas production from wheat straw stillage in UASB reactor. Applied Energy, 87(12), 3779-3783
9. Luo, G., Xie, L., Zou, Z., Wang, W., Zhou, Q. 2010b. Evaluation of pretreatment methods on mixed inoculum for both batch and continuous thermophilic biohydrogen production from cassava stillage. Bioresource Technolog,y 101(3), 959-964
10. Luo, G., Xie, L., Zou, Z., Wang, W., Zhou, Q., Wang, J.Y. 2010a. Fermentative hydrogen production from cassava stillage by mixed anaerobic microflora: Effects of temperature and pH. Applied Energy Volume, 87(12), 3710-3717.
11. Melamane X.L., Strong P.J., Burgess J.E. 2007. Treatment of wine distillery wastewater: a review with emphasis on anaerobic membrane reactors. South African Journal of Enology and Viticulture, 28(1), 25-36.
12. Moletta R. 2005. Winery and distillery wastewater treatment by anaerobic digestion. Water Science And Technology, 51(1), 137–144.
13. Nasr, N., Elbeshbishy, E., Hafez, H., Nakhla, G., El Naggar, M.H., 2011. Bio-hydrogen production from thin stillage using conventional and acclimatized anaerobic digester sludge. International Journal of Hydrogen Energy, 36(20),12761-12769.

14. Rada, E.C., Ragazzi, M. 2008. Critical analysis of PCDD/F emissions from anaerobic digestion, Water Science And Technology, 58.9, 1721-1725.

15. Rada, E.C., Ragazzi, M., Fiori, L., Antolini, D. 2009. Bio-drying of grape marc and other biomass: a comparison. Water Science And Technology, 60(4), 1065-1070.

16. Tokumoto, H., Tanaka, M. 2012. Novel anaerobic digestion induced by bacterial components for value-added byproducts from high-loading glycerol. Bioresource Technology 107, 327-332.

17. Vlissidis, A., Zouboulis, A.I., 1993. Thermophilic anaerobic digestion of alcohol distillery wastewaters. Bioresource Technology, 43, 131–140.

18. Wang, W., Xie, L., Chen, J., Luo, G., Zhou, Q. 2011. Biohydrogen and methane production by co-digestion of cassava stillage and excess sludge under thermophilic condition. Bioresource Technology, 102(4), 3833-3839.

19. Wolmarans, B., de Villiers, G.H., 2002. Start-up of a UASB effluent treatment plant on distillery wastewater. Water SA, 28, 63–68.

There are several supplemental files that are not available in this version of the article. To view this additional information, please use the citation on the first page of this chapter.

PART VI

FOOD PROCESSING AND PACKAGING

CHAPTER 11

Integrated Production and Treatment Biotech-Process for Sustainable Management of Food Processing Waste Streams

BO JIN

Food processing industry generates approximately 45% of the total organic pollution as wastewater and solid wastes. These organic pollutants pose increasing disposal and environmental challenges. The treatment and disposal of the organic wastes require many successive and costly treatment processes. These organic pollutants contribute high organic loading in organic carbon and nutrient sources. The processing effluents and solid wastes from food industries mainly contain carbohydrate organics such as sugars, starch and cellulose. They are biodegradable materials and naturally rich in nutrients, making them ideal substrates for microbial production [1,2]. Most of the existing treatment systems for the food processing wastes worldwide, however, are of old-fashioned processes and cause large

Integrated Production and Treatment Biotech-Process for Sustainable Management of Food Processing Waste Streams. © Jin B. AIMS Bioengineering *1,2 (2014). doi: 10.3934/bioeng.2014.2.88. Licensed under a Creative Commons Attribution 4.0 International License, http://creativecommons.org/licenses/by/4.0/.*

losses of valuable nutrient and carbon resources. Considering increasing global concerns due to greenhouse gas emission and resource crisis, there is a general agreement that environmental protection can only be achieved by integrating a general environmental awareness into a company's business functions, making the carbohydrate wastes as renewable resources.

In recent decades, conversion of waste materials into bulk chemicals and clean energy has become an attractive topic for applied research and technology development around the world. Bioconversion of carbohydrate wastes to valuable products is receiving increased attention in view of the fact that these wastes represent possible and utilizable resources for producing market products [3]. Growing research and technology development activities have been offered towards sustainable utilization and value-addition of carbohydrate wastes, such as sugar cane bagasse and waste starch as useful carbon and nutrient sources. A number of biotechnological processes have been developed that utilise these waste materials for the production of bulk chemicals and fine products such as ethanol, single cell protein, mushroom, enzymes, organic acids, amino acids, biological active secondary metabolites [1,2,3,4], as summarised in Figure 1. This paves a promising way for recycling of resources to become an integral activity in industry to ensure economical and ecological sustainability.

Biotechnological production from carbohydrate-wastes associated with their treatment requires an integrated engineering process strategy with concerns of utilization and treatment of the waste streams. Recent studies have given substantial R&D efforts to the development of an environmentally and economically sustainable integrated biotechnological process, so called "production treatment biotech-process" (PTB). The alternative PTB technology is to use the waste streams as bioconversion media to produce bulk products, while treating the carbohydrate waste streams [1,2,4]. The PTB engineering strategy is able to deliver an innovative "green cycle" technology, from which the value-added products, including those commonly used materials in food processing industries (organic acids and enzymes), can be produced from processing wastes, and can be served as feedstock for the industry, while reducing organic loadings. These advanced integrated technologies will give a significant contribution to update the traditional technologies of wastewater treatment and biological nutrient removal processes. Recent researches focused on developing and using technological tools of genetic and metabolic engineering and bioprocessing techniques in order to increase the production yield and the cost-efficacy of waste treatment [1,4,6]. There have been great advances in fundamental research into biochemical and chemical processing, biotechnological techniques and the genetic construction of high-performance industrial microorganisms with functional biochemical reaction capabilities in an industrial process [2,3,7].

The major limitations for the development towards application of the integrated PTB technologies with respect to economic and technique issues are (1) high costs for physical (steam expulsion) and chemical (alkali/acid hydrolysis, oxidation) processes employed for pretreatment of the raw materials, (2) long fermentation retention time, and (3) low efficiency for waste treatment [1,3]. The conventional industrial process for biological production from carbohydrate materials requires pretreatment by gelatinisation and liquefaction, which is carried out at high temperature of 90-130°C for 15 min followed by enzymatic saccharisation to glucose and subsequent conversion of glucose to organic acids by fermentation [8,9]. However, this two steps process involving consecutive enzymatic hydrolysis and microbial fermentation makes it economically unattractive. Alternatively, fermentation can be conducted simultaneously with the presence of enzymes known as 'Simultaneous Saccharification and Fermentation' (SSF) [4,7,10]. The SSF technology eliminates the need for complete hydrolysis step prior to the fermentation step. It is expected that the hydrolysis of polysaccharide and fermentation of glucose into biochemicals can be carried out simultaneously in the SSF process [10]. In the SSF process biochemical reactions in term of enzymatic hydrolysis, cell growth and metabolic production may occur simultaneously and or stepwise. The emerging tools of genetic engineering, however, have also stimulated the construction of microbial production strains for the direct synthesis of new metabolites within a SSF process model [9,10].

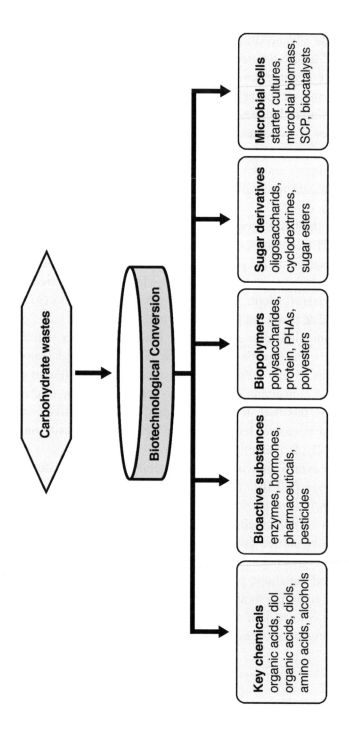

FIGURE 1: Target products from bioconversion of carbohydrate wastes to value-added products.

The enzymes as biocatalysts play key roles in the integrated biotech-process through substrate hydrolysis and metabolic formation. However, applications of these enzymes in an industrial process are limited by their instability and non-reusability. Enzyme immobilization represents the most promising approach to improving stability, loading, activity, and speciality of the enzymes. Nano-biocatalysis, in which enzymes are incorporated into nano-structured materials, has emerged as a rapidly growing area. Recent development in nanotechnology has provided a wealth of diverse nano-scale folds that could potentially support enzyme immobilisation. Nano-structures exhibit high specific surface areas and provide efficient manipulators, creating nano-scale microenvironment benefiting the immobilization of multienzymes and their interfacial reactions. They promise exciting advantages for improving enzyme stability and capability, and biochemical performances. The high specific surface area of the nano-immobilisers has been a principal driving force for studying and developing biocatalysts. Furthermore, enzyme immobilization using nano-structure carries allows for a significant increase in life cycles of the biocatalyst for its reuse, hence reducing the cost of the biocatalytic process [11]. The nanoscale-material provides a versatile new technology for enzyme immobilization with several inherent advantages, including low cost, rapid immobilization and reaction, similarity of nanosize, mild conversion condition, robust activities, mobility, high loading, minimum diffusional limitation, self-assembly and stability. The mobility, confining effects, solution behaviours and interfacial properties of nanoscale materials can introduce unique properties into nano-biocatalyst systems, making it possible to develop a revolutionary class of biocatalysts that differs from conventional immobilized enzymes in terms of preparation, catalytic efficiency and application potential [12].

Integrated production and treatment technology is an innovative and sustainable R&D practice. Incorporation and development of recent advances in metabolic production process and biological wastewater treatment technology will lead to improved genetic and biochemical engineering, and a deep practical insight of the integrated production treatment process. The development of integrated PTB technology will provide R&D opportunity to apply and develop processing technologies with integration of metabolic production, waste treatment and genetic engineering.

The "green cycle" bioprocess will improve the old-fashioned waste treatment and recovery technologies, and take an important role for the cleaner production strategy in industries.

The integrated PTB engineering strategy is of today's important challenges for the sustainable production of renewable resources from waste organic materials. The advanced biotechnology and nanotechnology have significantly promoted the development of the biocatalysts and bioconversion technologies for the production of bioenergy and biomaterials from renewable sources, especially from industrial and agricultural organic waste streams.

REFERENCES

1. Sankar S, Khanal SK, Jasti N, et al. (2010) Use of Filamentous Fungi for Wastewater Treatment and Production of High Value Fungal Byproducts: A Review. Critical Reviews in Environ. Sci Technol 40: 1-49.
2. Li WW, Yu HQ (2011) From wastewater to bioenergy and biochemicals via two-stage bioconversion processes: A future paradigm. Biotechnol Adv 29: 972-982.
3. Van Dyk JS, Pletschke BI (2012) A review of lignocellulose bioconversion using enzymatic hydrolysis and synergistic cooperation between enzymes—Factors affecting enzymes, conversion and synergy. Biotechnol Adv 30: 1458-1480.
4. Jin B, Yu Q, van Leeuwen J (2001) A bioprocessing mode for fungal biomass protein production and wastewater treatment using external air-lift bioreactor. J Chem Technol Biotechnol 76: 1041-1048.
5. Mohanram S, Amat D, Choudhary J, et al. (2013) Novel perspectives for evolving enzyme cocktails for lignocellulose hydrolysis in biorefineries. Sustainable Chem Process 1: 1-15
6. Kaur B, Oberoi HS, Chadha BS (2014) Enhanced cellulase producing mutants developed from heterokaryotic Aspergillus strain. Bioresour Technol 156: 100-107.
7. Clark JH, Luque R, Matahru AS (2012) Green chemistry, biofuels and biorefinery. Annu Rev Chem Biomol Eng 3: 183-207.
8. Alvira P, Tomas-Pejo E, Ballesteros M, et al. (2010) Pretreatment technologies for an efficient bioethanol production process based on enzymatic hydrolysis: a review. Bioresour Technol 101: 4851-61.
9. Suhardi VSH, Prasai B, Samaha D, et al. (2013) Combined biological and chemical pretreatment method for lignocellulosic ethanol production from energy cane. Renew Bioresour Available from: http://www.hoajonline.com/renewablebioresources/2052-6237/1/1.
10. Banerjee G, Scott-Craig JS, Walton JD (2010) Improving enzymes for biomass conversion: a basic research perspective. Bioenerg Res 3: 82-92.

11. Misson M, Zhang H, Jin B (2014) Nanobiocatalyst advancements and bioprocessing applications. Interface. Available from: http://rsif.royalsocietypublishing.org/content/12/102/20140891.

12. Betancor L, Luckarift HR (2008) Bioinspired enzyme encapsulation for biocatalysis. Trends Biotechnol 26: 566-572.

PART VII

CONCLUDING IMPLICATIONS

CHAPTER 12

Are the Dietary Guidelines for Meat, Fat, Fruit and Vegetable Consumption Appropriate for Environmental Sustainability? A Review of the Literature

CHRISTIAN JOHN REYNOLDS, JONATHAN DAVID BUCKLEY, PHILIP WEINSTEIN, AND JOHN BOLAND

12.1 INTRODUCTION

Food consumption contributes an estimated 20% to 30% of the total adverse environmental impact in the Western world [1,2], being linked to soil, air, and water pollution and loss of biodiversity [3,4,5,6,7,8,9,10,11, 12,13,14,15,16]. Despite this having been recognised for some time, the idea of altering diet to increase environmental sustainability is a relatively new concept that until the last decade had little real-life implementation [3,4,17,18].

Are the Dietary Guidelines for Meat, Fat, Fruit and Vegetable Consumption Appropriate for Environmental Sustainability? A Review of the Literature. © Reynolds CJ, Buckley JD, Weinstein P, and Boland J. Nutrients **6,**6 (2014), doi:10.3390/nu6062251. Licensed under a Creative Commons Attribution 3.0 Unported License, http://creativecommons.org/licenses/by/3.0/.

Currently, dietary guidelines in most jurisdictions are mainly used to promote healthy eating to prevent chronic disease [19,20], with environmental, economic, or social impacts of diet considered to be externalities. Typically, as seen in the global dietary guidelines discussed by the World Health Organization (WHO) and the United Nations Food and Agriculture Organization (FAO) [21], the dialogue on environmental benefits is tempered by the focus on health, and the need to provide practical advice that people can follow. Environmental considerations, if mentioned, are relegated to the appendices—as in the current Australian Dietary guidelines [22,23]. Rare exceptions to this trend are the recent publications by the Health Council of the Netherlands [24] and the Nordic Council of Ministers [25] which discuss a healthy diet from an ecological perspective.

Globally there is much debate over what constitutes a healthy diet, how to optimise diet, and how to present this information to the general population [26,27,28,29,30,31,32,33]. There are national and international "healthy" portion sizes, and recommended daily allowances of differing foodstuffs based upon caloric content, cultural, historic and economic factors. Yet many studies express difficulty in finding an individual eating in perfect concordance with global dietary guidelines [21,34,35,36].

Since the 1960s there has been a marked increase in the variety of dietary guidelines published [37,38,39]. Concurrently, the global diet has shifted due to the rising global average income, and greater access to cheap, highly processed foodstuffs and animal products, resulting in increasing rates of obesity and chronic disease [40]. The intensification of publication and debate over recommended diets can be seen, in part, as a reaction to this changing global diet and the adverse impacts on health. Yet, as indicated above, discussion around the environmental impacts of recommended dietary guidelines is now only emerging [41].

In this paper we focus on dietary guidelines, excluding nutrient guidelines (such as [42]). The sheer variety of foods that an individual can choose from to obtain the recommended daily nutrients, results in greater complexity with respect to the associated environmental impacts. Due to the complexity we believe that nutrient guidelines are deserving of their own analysis.

In this paper we review the current environmental impact assessment and life cycle analysis (LCA) literature around the environmental impacts of dietary recommendations, focusing on collating the environmental evidence behind three pieces of dietary advice that are debated in current environmental impact assessment and LCA literature, and also presented in the WHO guidelines: reducing the consumption of fat, reducing the consumption of meat-based protein and animal-based foods, and increasing the consumption of fruit and vegetables [21,32,33]. Environmental impacts of dietary advice presented in WHO guidelines to reduce sugar consumption have not been evaluated in the LCA literature and was therefore unable to be included in this review. The lack of assessment of the impact of this guideline indicates a gap in the literature that should be addressed in future studies.

12.2 REDUCING FATS

Since the 1960s the average global daily consumption of fat has increased by 20 g per person (27%) [21]. This rise in fat consumption is proposed to be due to an increase in the availability and consumption of cheaper energy-dense, high-fat, nutrients-poor food stuffs [43,44,45,46] such as processed snacks, caloric beverages, fast foods, and edible oils and spreadable fats. The increase in fat intake, and associated increase in energy intake, has been blamed for contributing to an epidemic increase in overweight and obesity and associated health conditions, with 1.5 billion people classified as overweight and over 500 million as obese [40,47].

Fats can be understood to be an independent category of food, as well as a nutritional component of a broad range of foodstuffs. Thus, fats can be directly consumed as edible oils and spreads, or indirectly consumed in food sources such as dairy, meat, etc. This dual nature of the dietary availability of fats has meant that dietary and nutrient advice has overlapped in discussing healthy fat consumption levels. As a result most dietary guidelines provide recommendations for total fat consumption, rather than specific recommendations for consumption of fat as edible oils and spreadable fats. The current WHO guidelines recommend that 15%–

30% of dietary energy be supplied from fats. However, the actual amount of dietary energy derived from fats is country dependent, with the figure for developed nations being around 20%–40% [21,48,49]. This recommendation by the WHO to limit fat consumption is based on recommendations aimed at improving health rather than environmental impacts, and relates to total intake of fat from all sources, including not only foods from the food category of edible fats and oils, but also fat contained as part of the nutrient profile of other foods. In terms of environmental impacts however, because of the assessment methodologies used, it has to date only been possible to estimate the environmental impacts of edible fats and oils as a food category and not to evaluate the independent environmental effects of fats that form part of the composition of other food categories. For example, meat contains fat as part of its nutritional profile, but separate analysis cannot currently be undertaken to estimate the environmental impact of the fat content of the meat independently from the protein and other nutrients as the current methodologies available do not permit this level of analysis. Thus, at present analysis of the environmental impacts of fat is limited to analysis of the food category of edible fats and oils.

Vieux et al. [50] examined the greenhouse gas effects of reducing the consumption of energy-dense, high-fat, nutirent-poor food stuffs, and found that the food category of edible fats contributed 7% of daily diet-associated greenhouse gas emissions compared to fruit and vegetables at 9%, or meat at 27%. Considering that the French diet sources over 40% of its energy from fat-type products [49], it can be understood that edible fats do not have a large (nor proportional) environmental impact when compared to fruit, vegetables or meat products. In a subsequent study by these same authors, which also evaluated total dietary fat intake (i.e., nutrient analysis), it was reported that while total dietary fat intake was higher in lower nutritional quality diets, these diets were associated with lower greenhouse gas emissions [51], but this latter study was unable to identify the specific contribution of dietary fat to greenhouse gas emissions and the lower greenhouse gas emissions may have been related to nutrients other than fat. One strength of these studies was that they evaluated self-selected diets from a random sample of the French population, and therefore reflect the environmental impacts of the actual diversity of food consumption patterns, but a limitation was that they examined the environmental impacts of different food categories rather than the impacts of individual dietary nutrients. Additional research should examine the proportion of environmental consequences from the intake of indirect fats (as nutrients) as opposed to the food category of fats and oils.

Despite the apparently low environmental impact of fat consumption, it has been proposed that fat intake be reduced to improve health, with a possible mechanism to facilitate this being the introduction of a "fat" tax. However, it has been proposed that such a tax would represent an economic device to raise revenue rather than alter diet [43,45,46,52,53,54,55,56,57]. Conversely, it has been further suggested that fat taxes could be used to increase prices to reflect the actual social cost of food, including the cost of ameliorating environmental impacts [58,59,60,61]. A common criticism of "fat" taxes is that they are regressive—with low-income households being forced pay a greater percentage of their income than higher income households. Furthermore, the environmental impacts of a fat tax have not been well examined, with only Friel et al. [62] providing some discussion on the merits of fat taxes to reduce consumption of GHG intensive goods. In particular, Friel et al. [62] discussed that fat-taxes—though useful— should only be part of a behaviour change tool set, and could also be used to "link health and climate-change agendas".

Thus, the reduction of fat consumption via price mechanisms may produce some monetary and health benefits. However, the environmental impact of a reduction in fat consumption is unclear as methods are only available to model the impact of reduced intake of edible fats and oils as a food category and not the impacts of fats that form a nutritional component of other food groups. However, a reduction of fat consumption via the eating of low fat (or fat removed) foods could result in the removed fat becoming food waste if not used by other industries [63,64,65]. In turn this waste could produce problematic environmental consequences. Further study of the indirect environmental impacts of a low fat diet is required.

12.3 REDUCING MEAT AND ANIMAL-BASED FOODS

Since the 1960s, the consumption of animal-based foods has risen throughout the world at the expense of consumption of non-animal-based staple foods such as grains, pulses, and fruits and vegetables [6]. This is due to increased production efficiency of the meat and dairy industries [66,67,68,69,70], higher standards of living and a rising global average income with an increasing demand for meat [71,72,73]. This is most evident in China with total meat consumption increasing 165% since 1990, while in Asia as a whole it has increased 30-fold since 1961 [74]. However, there is some evidence that meat consumption in Asia may have peaked, and that these countries may now not be following the developed world's consumption pattern for more meat [71,73,74].

The consumption of meat and animal products offers essential (micro) nutritional security to many who would be otherwise food insecure [75,76,77]. However, excessive consumption of meat and animal products in some countries, and in some social classes within countries, can lead to excessive intakes of fat (nutrients), which can impact adversely on health [78]. This has led to the recommendation in some dietary guidelines [22] to limit meat consumption, in particular processed meat and, for men to reduce their intake of red meat. Even with such recommendations, the FAO is projecting a global yearly consumption of 45 kg of meat and 95 kg of dairy per person by the year 2030 [21]. Though this is below the 1997/9 average yearly meat and milk consumption of both industrialised (88 kg, 212 kg) and transition (46 kg, 159 kg) economies [71], is still above levels that many consider to be sustainable [74,79,80].

The environmental impact of meat and animal product consumption has been the topic of some investigation [6,67,68,76,79,80,81,82,83,84,85,86]. It has been found that meat-centric meals generate on average nine times higher greenhouse gas emissions than plant-based equivalents [5], while specific meat-based products such as beef and cheese cause 10–20 times more environmental impact [87,88,89]. An animal-based diet requires 2.5–5.0 times the energy inputs [7,90,91,92,93], 2–3 times the water, 13 times the fertilizer, and 1.4 times the pesticide use per calorie produced compared with a plant-based diet [7,93,94]. In European life cycle assessment studies, because of the relatively high meat intake in the typical diet, meat-free scenarios were between 18% and 31% lower in greenhouse gas emissions than the average diet [17,41].

Though there is literature suggesting an environmentally friendly diet can be achieved with meat and dairy products present [28], there have been many arguments mounted against meat dominant diets [5,93,95,96,97,98]. If animal-based foods are to be part of the diet, the selection of the least environmentally damaging foods is crucial. McMichael et al. [98] modelled a working global yearly meat intake target of no more than 32 kg per person with no more than 18 kg per year coming from red meat from ruminants (i.e., cattle, sheep, goats, etc.). This is below the projected intake of 45 kg of meat for a meat-reduced diet cited in the aforementioned WHO/FAO meat consumption projections.

Besides selecting non-ruminant animals, another way to minimise the impact of animal products is to use farming practices that are suitable to the land type, and select less environmentally damaging feed and fodder [17]. These farming practices (lot-fed compared to grass-fed) can result in pronounced differences of environmental impacts. Studies [87,99,100] have shown this variance to be dependent on the type and geography of the farmed land and the differences can be minimal [101]—grass feeding having a potentially greater environmental impact than lot feeding in arid areas, while in developed nations with temperate climates, lot feeding can have greater impacts depending on the production systems [86].

The larger contribution to the environmental impact of animals that can be altered is the feed used [102,103,104]. Currently there is a large dependency on cereals and legumes (such as wheat, corn and soy) for animal feed with 37% of global cereal production being fed to animals [76]. Traditionally farming of these cereals has been resource intensive with a sizeable environmental footprint [105,106]. Regardless of the sustainable intensification techniques that are now being implemented, demand for these cereals as animal feed (along with bio fuel production) is currently growing, resulting in global deforestation and biodiversity threats [8]. Switching to alternative sources of animal feed that are less resource intensive might therefore provide a viable method for reducing the environmental impact of animal-based foods. More potentially sustainable alternatives include by-products from other agricultural sectors (such as molasses cake, brewers' grains, vegetable residues and rice husks) [76]. The current level of production from by-product feedstock is limited, despite Fadel [107] finding that there was theoretically enough nutritional content available to provide production for 80% of global milk consumption in 1993. However, this study included processed soymeal as a feed, which Garnett [76] indicates is not a by-product per se, nor can soy meal be claimed to be resource efficient, because industrial soy farming can have many negative environmental consequences [104,108,109].

Fish and seafood consumption provides animal protein from a non-red meat source. Currently, fish and seafood provides more than 4 billion people with approximately 15% of their intake of animal protein, which equates to a global yearly consumption of approximately 18.6 kg of fish per person [21,110]. This consumption of fish is above the level of population growth, with 57% of global fish stocks now fully exploited (i.e., at or very close to their maximum sustainable production) and 30% overexploited [110]. To meet demand and to combat the problems of over fishing wild-caught fish stocks there has been a marked increase in global aquaculture (with much occurring in Asia and Sub-Saharan Africa [111,112,113,114,115]), and a recent World Bank report [116] stated that by 2030, aquaculture will provide close to two thirds of global food fish consumption (186 million tons). With the increasing use of aquaculture, Merino et al. [117] have determined that the fish demand by 2050 will be met, but only if fish resources are managed sustainably and the animal feeds industry reduces its reliance on wild fish. While increasing aquaculture may assist in preventing depletion of wild fish stocks, both wild-caught and aquafarmed fish have substantial environmental impacts, with fish protein being up to 14 times more energy-intensive to produce than a vegetable equivalent [93].

From these statistics, the case can easily be made that reducing the intake of animal protein (including from fish) and dairy foods in the global diet would potentially have considerable impact on reducing environmental effects. However, this is likely to be unpalatable to much of the global population for many cultural, nutritional, and economic reasons [28,118]. Nevertheless, from an environmental perspective the dietary advice to reduce animal based foods is most welcome.

12.4 INCREASING FRUIT AND VEGETABLE INTAKE

Fruits and vegetables play a key role in providing a diverse and nutritious diet, with studies showing that adequate consumption of fruits and vegetables reduces the risk of chronic disease [119,120,121]. However, unlike red meat, the global consumption of fruits and vegetables has persistently been below recommended guidelines [122], with over 77% of men and women in low- and middle-income countries consuming less than the WHO's minimum recommended 400 g per capita per day of fruits and vegetables. Consumption of fruits and vegetables for many high income countries is also lower than the WHO's minimum recommended volumes [21,33,34,35,123].

The environmental impact of fruits and vegetables varies greatly according to the individual type and production method [124,125,126,127,128]. Thus, it is more useful to contrast typical diets with diets high in vegetables and fruit, or high in animal-based foods [5,98,129,130]. In these dietary comparisons it has been found that greenhouse gas emissions with diets high in vegetables and fruit are lower than typical diets or diets high in animal-based foods. A number of studies have also reported that vegetarian diets are more environmentally friendly than other dietary patterns [7,93,94,131]. Furthermore, Baumann [132] found that vegan diets produced 23% less greenhouse gas emissions than the average vegetarian diet. However, it should be noted that from a food security and diet perspective, vegan and vegetarian diets, though lower in environmental impacts, have nutritional risks [133]. Specifically, there is greater potential for the insufficient intake of certain micronutrients [41,88,134]. This matters most when diet is of a monotonous limited selection, when even a small animal-based food intake could make a critical difference to micronutrient intake [75,76,77]. Thus the advice for a diet high in fruit and vegetables, with some meat products, has some merit from the food security viewpoint as well as opposed to vegetarian or vegan diets.

While the majority of evidence suggests that an increased intake of fruit and vegetables will reduce environmental impact, there is a small (but growing) literature that suggests a diet low in meat and high in fruits and vegetables is not always low in environmental impact [135]. This is because, in some cases, the quantity of vegetable substitutes eaten to replace animal proteins can contribute similar levels of environmental impacts [50,51], due to the increased quantities of cereals and vegetables for human consumption only slightly outweighing the corresponding decline in the land, water, and resources required to grow feed-cereal previously destined for animals [49]. There needs to be additional modelling to test these claims.

12.5 CONCLUSIONS

There are a myriad of possibly sustainable diets, with the components of each part of a diet contributing different volumes of environmental impacts [28]. In this paper we have examined three of the most common pieces of advice found in dietary guidelines. We found evidence of environmental benefits from reducing animal product intake and increasing fruit and vegetable consumption. However, there is also a small (but growing) section of the literature that suggests a diet low in meat and high in fruits and vegetables is not always the most environmentally friendly [135]. Further study is required to examine the veracity and suitability of these claims for the various global diets.

We found little research into the environmental impact of reducing fat in the diet. The most recent study to examine the environmental impact of direct edible fats found that fat currently accounts for less GHG emissions than vegetables, while contributing a larger share of dietary energy. This finding gives weight to the argument that the diets lowest in GHG emissions may not be lowest in fat [51]. Further investigation is needed into the environmental impacts of both direct and indirect fats within contemporary global diets.

Importantly, there is clear evidence that the majority of the global population does not adhere to dietary advice. Our review suggests that further investigation into the environmental benefits of following dietary guidelines in comparison to contemporary reported dietary habits is required. Such evidence would give more strength to the argument for adopting recommended dietary guidelines.

REFERENCES

1.   Tukker, A.; Jansen, B. Environmental Impacts of Products: A Detailed Review of Studies. J. Ind. Ecol. 2006, 10, 159–182.

2. Foley, J.A.; Ramankutty, N.; Brauman, K.A.; Cassidy, E.S.; Gerber, J.S.; Johnston, M.; Mueller, N.D.; O'Connell, C.; Ray, D.K.; West, P.C.; et al. Solutions for a cultivated planet. Nature 2011, 478, 337–342.

3. Gussow, J.D.; Clancy, K.L. Dietary guidelines for sustainability. J. Nutr. Educ. 1986, 18, 1–5.

4. Goodland, R. Environmental sustainability in agriculture: Diet matters. Ecol. Econ. 1997, 23, 189–200.

5. Carlsson-Kanyama, A. Climate change and dietary choices—How can emissions of greenhouse gases from food consumption be reduced? Food Policy 1998, 23, 277–293.

6. Steinfeld, H.; Gerber, P.; Wassenaar, T.; Castel, V.; Rosales, M.; de Haan, C. Livestock's Long Shadow; FAO: Rome, Italy, 2006.

7. Marlow, H.J.; Hayes, W.K.; Soret, S.; Carter, R.L.; Schwab, E.R.; Sabaté, J. Diet and the environment: Does what you eat matter? Am. J. Clin. Nutr. 2009, 89, 1699S–1703S.

8. Lenzen, M.; Moran, D.; Kanemoto, K.; Foran, B.; Lobefaro, L.; Geschke, A. International trade drives biodiversity threats in developing nations. Nature 2012, 486, 109–112.

9. Pinstrup-Andersen, P.; Pandya-Lorch, R. Food security and sustainable use of natural resources: A 2020 vision. Ecol. Econ. 1998, 26, 1–10.

10. Lorek, S.; Spangenberg, J.H. Indicators for environmentally sustainable household consumption. Int. J. Sustain. Dev. 2001, 4, 101–120.

11. Tukker, A.; Goldbohm, R.A.; de Koningb, A.; Verheijden, M.; Kleijn, R.; Wolf, O.; Pérez-Domínguez, I.; Rueda-Cantuche, J.M. Environmental impacts of changes to healthier diets in Europe. Ecol. Econ. 2011, 70, 1776–1788.

12. Jackson, T.; Papathanasopoulou, E. Luxury or 'lock-in'? An exploration of unsustainable consumption in the UK: 1968 to 2000. Ecol. Econ. 2008, 68, 80–95.

13. Druckman, A.; Jackson, T. The carbon footprint of UK households 1990–2004: A socio-economically disaggregated, quasi-multi-regional input-output model. Ecol. Econ. 2009, 68, 2066–2077.

14. Reisch, L.; Scholl, G.; Eberle, U. Discussion Paper 1 on Sustainable Food Consumption; CORPUS: Brussels, Belgium, 2010.

15. Csutora, M. One More Awareness Gap? The Behaviour-Impact Gap Problem. J. Consum. Policy 2012, 35, 145–163.

16. Vetőné Mózner, Z.; Csutora, M. Designing lifestyle-specific food policies based on nutritional requirements and ecological footprints. Sustain. Sci. Pract. Policy 2013, 9, 48–59.

17. Berners-Lee, M.; Hoolohan, C.; Cammack, H.; Hewitt, C.N. The relative greenhouse gas impacts of realistic dietary choices. Energy Policy 2012, 43, 184–190.

18. Joyce, A.; Dixon, S.; Comfort, J.; Hallett, J. Reducing the Environmental Impact of Dietary Choice: Perspectives from a Behavioural and Social Change Approach. J. Environ. Public Health 2012, 2012, 7.

19. Nishida, C.; Uauy, R.; Kumanyika, S.; Shetty, P. The joint WHO/FAO expert consultation on diet, nutrition and the prevention of chronic diseases: Process, product and policy implications. Public Health Nutr. 2004, 7, 245–250.

20. WHO and FAO. Preparation and Use of Food-Based Dietary Guidelines; World Health Organisation: Geneva, Switzerland, 1998.
21. WHO and FAO. Diet, Nutrition and the Prevention of Chronic Diseases; WHO Technical Report Series, No. 916; World Health Organisation: Geneva, Switzerland, 2003.
22. National Health and Medical Research Council. EAT FOR HEALTH Australian Dietary Guidelines Providing the Scientific Evidence for Healthier Australian Diets; National Health and Medical Research Council: Canberra, Australia, 2013.
23. National Health and Medical Research Council. A Review of Evidence to Address Targeted Questions to Inform the Revision of the Australian Dietary Guidelines; Australian Government Department of Health and Ageing: Canberra, Australia, 2011.
24. Health Council of the Netherlands. Guidelines for a Healthy Diet: The Ecological Perspective; Health Council of the Netherlands: The Hague, The Netherlands, 2011.
25. Ministers, N.C.O. Nordic Nutrition Recommendations 2012; Nord: Copenhagen, Denmark, 2014.
26. Rozenbergs, V.; Skrupskis, I.; Skrupska, D.; Rozenberga, E. Food allowance optimization model. In Proceedings of the International Scientific Conference 'Rural En, Jelgava, Latvia, 2013; p. 347.
27. Wilson, N.; Nghiem, N.; Mhurchu, C.N.; Eyles, H.; Baker, M.G.; Blakely, T. Foods and Dietary Patterns That Are Healthy, Low-Cost, and Environmentally Sustainable: A Case Study of Optimization Modeling for New Zealand. PLoS One 2013, 8, e59648.
28. Macdiarmid, J.I.; Kyle, J.; Horgan, G.W.; Loe, J.; Fyfe, C.; Johnstone, A.; McNeill, G. Sustainable diets for the future: Can we contribute to reducing greenhouse gas emissions by eating a healthy diet? Am. J. Clin. Nutr. 2012, 96, 632–639.
29. Shrapnel, B.; Baghurst, K. Lack of nutritional equivalence in the 'meats and alternatives' group of the Australian guide to healthy eating. Nutr. Diet. 2007, 64, 254–260.
30. Rangan, A.M.; Schindeler, S.; Hector, D.J.; Gill, T.P. Assessment of typical food portion sizes consumed among Australian adults. Nutr. Diet. 2009, 66, 227–233.
31. Nestle, M. Food Politics: How the Food Industry Influences Nutrition and Health 2007; University of California Press: London, UK.
32. WHO and FAO. Human Energy Requirements: Report of a Joint FAO/WHO/UNU Expert Consultation; United Nations University: Rome, Italy, 2001.
33. WHO. FAO/WHO Technical Consulation on National Food-Based Dietary Guidelines; WHO: Geneva, Switzerland, 2006.
34. McNaughton, S.A.; Ball, K.; Crawford, D.; Mishra, G.D. An index of diet and eating patterns is a valid measure of diet quality in an Australian population. J. Nutr. 2008, 138, 86–93.
35. Folsom, A.R.; Parker, E.D.; Harnack, L.J. Degree of concordance with DASH diet guidelines and incidence of hypertension and fatal cardiovascular disease. Am. J. Hypertens. 2007, 20, 225–232.
36. MacLennan, R.; Zhang, A. Cuisine: The concept and its health and nutrition implications-global. Asia Pac. J. Clin. Nutr. 2004, 13, 131–135.
37. Gifford, K.D. Dietary fats, eating guides, and public policy: History, critique, and recommendations. Am. J. Med. 2002, 113, 89–106.

38. Aranceta, J.; Pérez-Rodrigo, C. Recommended dietary reference intakes, nutritional goals and dietary guidelines for fat and fatty acids: A systematic review. Br. J. Nutr. 2012, 107, S8–S22.
39. Truswell, A.S. Evolution of dietary recommendations, goals, and guidelines. Am. J. Clin. Nutr. 1987, 45, 1060–1072.
40. Swinburn, B.A.; Sacks, G.; Hall, K.D.; McPherson, K.; Finegood, D.T.; Moodie, M.L.; Gortmaker, S.L. The global obesity pandemic: Shaped by global drivers and local environments. Lancet 2011, 378, 804–814.
41. Meier, T.; Christen, O. Environmental impacts of dietary recommendations and dietary styles: Germany as an example. Environ. Sci. Technol. 2012, 47, 877–888.
42. Australian Government; Department of Health and Ageing Australia; National Health and Medical Research Council. Nutrient Reference Values for Australia and New Zealand Including Recommended Dietary Intakes. Commonwealth of Australia: Canberra, Australia, 2006.
43. Andrieu, E.; Darmon, N.; Drewnowski, A. Low-cost diets: More energy, fewer nutrients. Eur. J. Clin. Nutr. 2005, 60, 434–436.
44. Drewnowski, A.; Darmon, N.; Briend, A. Replacing fats and sweets with vegetables and fruits—A question of cost. Am. J. Public Health 2004, 94, 1555–1559.
45. Darmon, N.; Briend, A.; Drewnowski, A. Energy-dense diets are associated with lower diet costs: A community study of French adults. Public Health Nutr. 2004, 7, 21–27.
46. Darmon, N.; Ferguson, E.L.; Briend, A. A cost constraint alone has adverse effects on food selection and nutrient density: An analysis of human diets by linear programming. J. Nutr. 2002, 132, 3764–3771.
47. Garnett, T. Food sustainability: Problems, perspectives and solutions. Proc. Nutr. Soc. 2013, 72, 29–39.
48. Tataranni, P.A.; Ravussin, E. Effect of fat intake on energy balance. Ann. N. Y. Acad. Sci. 1997, 819, 37–43.
49. Srinivasan, C.S.; Irz, X.; Shankar, B. An assessment of the potential consumption impacts of WHO dietary norms in OECD countries. Food Policy 2006, 31, 53–77.
50. Vieux, F.; Darmon, N.; Touazi, D.; Soler, L.G. Greenhouse gas emissions of self-selected individual diets in France: Changing the diet structure or consuming less? Ecol. Econ. 2012, 75, 91–101.
51. Vieux, F.; Soler, L.-G.; Touazi, D.; Darmon, N. High nutritional quality is not associated with low greenhouse gas emissions in self-selected diets of French adults. Am. J. Clin. Nutr. 2013, 97, 569–583.
52. Caraher, M.; Cowburn, G. Taxing food: Implications for public health nutrition. Public Health Nutr. 2005, 8, 1242–1249.
53. Mytton, O.; Gray, A.; Rayner, M.; Rutter, H. Could targeted food taxes improve health? J. Epidemiol. Community Health 2007, 61, 689–694.
54. Bonnet, C. How to Set up an Effective Food Tax?; Comment on "Food Taxes: A New Holy Grail?". Int. J. Health Policy Manag. 2013, 1, 233–234.
55. Wilson, N.; Mansoor, O. Food pricing favours saturated fat consumption: Supermarket data. N. Z. Med. J. 2005, 118, U1338.
56. Wallace, J. Easy on the Oil: Policy Options for a Smaller Waistline and a Lighter Footprint; South Australian Department of Premier and Cabinet: Adelaide, Austra-

lia, 2009. Available online: www.brass.cf.ac.uk/uploads/Wallace_A70.pdf (accessed on 10 January 2014).

57. Chouinard, H.H.; Davis, D.; LaFrance, J.; Perloff, J. Fat taxes: Big money for small change. Forum Health Econ. Policy 2007, 10, 1071–1071.

58. Cash, S.B.; Sunding, D.L.; Zilberman, D. Fat taxes and thin subsidies: Prices, diet, and health outcomes. Acta Agric. Scand Sect. C 2005, 2, 167–174.

59. Vinnari, M.; Tapio, P. Sustainability of diets: From concepts to governance. Ecol. Econ. 2012, 74, 46–54.

60. Wirsenius, S.; Hedenus, F.; Mohlin, K. Greenhouse gas taxes on animal food products: Rationale, tax scheme and climate mitigation effects. Clim. Chang. 2011, 108, 159–184.

61. Lombardini, C.; Lankoski, L. Forced Choice Restriction in Promoting Sustainable Food Consumption: Intended and Unintended Effects of the Mandatory Vegetarian Day in Helsinki Schools. J. Consum. Policy 2013, 36, 159–178.

62. Friel, S.; Dangour, A.D.; Garnett, T.; Lock, K.; Chalabi, Z.; Roberts, I.; Butler, A.; Butler, C.D.; Waage, J.; McMichael, A.; et al. Public health benefits of strategies to reduce greenhouse-gas emissions: Food and agriculture. Lancet 2009, 374, 2016–2025.

63. Hall, K.D.; et al. The progressive increase of food waste in America and its environmental impact. PLoS One 2009, 4, e7940.

64. Weiss, T.J. Food Oils and Their Uses; Ellis Horwood Ltd.: Chichester, UK, 1983.

65. Gunstone, F.D. The Chemistry of Oils and Fats: Sources, Composition, Properties, and Uses; CRC Press: Oxford, UK, 2004.

66. Capper, J.L. The environmental impact of beef production in the United States: 1977 Compared with 2007. J. Anim. Sci. 2011, 89, 4249–4261.

67. Cederberg, C.; Mattsson, B. Life cycle assessment of milk production—A comparison of conventional and organic farming. J. Clean. Prod. 2000, 8, 49–60.

68. Cederberg, C.; Stadig, M. System expansion and allocation in life cycle assessment of milk and beef production. Int. J. Life Cycle Assess. 2003, 8, 350–356.

69. Thomassen, M.A.; Dalgaard, R.; Heijungs, R.; de Boer, I. Attributional and consequential LCA of milk production. Int. J. Life Cycle Assess. 2008, 13, 339–349.

70. Vellinga, T.V.; Gerber, P.; Opio, C. Greenhouse gas emissions from global dairy production. Sustain. Dairy Prod. 2013, 2013, 9–30.

71. Bruinsma, J. World Agriculture: Towards 2015/2030: An FAO Perspective; Earthscan: London, UK, 2003.

72. Hallström, E.; Börjesson, P. Meat-consumption statistics: Reliability and discrepancy. Sustain. Sci.Pract. Policy 2013, 9, 37–47.

73. Rivers Cole, J.; McCoskey, S. Does global meat consumption follow an environmental Kuznets curve. Sustain. Sci. Pract. Policy 2013, 9, 26–36.

74. Schwarzera, S.; Witta, R.; Zommers, Z. Growing Greenhouse Gas Emissions Due to Meat Production; UNEP GEAS: Nairobi, Kenya, 2012.

75. Neumann, C.; Harris, D.M.; Rogers, L.M. Contribution of animal source foods in improving diet quality and function in children in the developing world. Nutr. Res. 2002, 22, 193–220.

76. Garnett, T. Livestock-related greenhouse gas emissions: Impacts and options for policy makers. Environ. Sci. Policy 2009, 12, 491–503.

77. Buttriss, J.; Riley, H. Sustainable diets: Harnessing the nutrition agenda. Food Chem. 2013, 140, 402–407.
78. Hooper, L.; Summerbell, C.D.; Higgins, J.P.T.; Thompson, R.L.; Capps, N.E.; Smith, G.D.; Riemersma, R.A.; Ebrahim, S. Dietary fat intake and prevention of cardiovascular disease: Systematic review. BMJ 2001, 322, 757–763.
79. Heinrich Böll Foundation and Friends of the Earth Europe. MEAT ATLAS Facts and Figures about the Animals We Eat; Möller Druck: Ahrensfelde, Germany, 2014.
80. Ripple, W.J.; Smith, P.; Haberl, H.; Montzka, S.A.; McAlpine, C.; Boucher, D.H. Ruminants, climate change and climate policy. Nat. Clim. Chang. 2014, 4, 2–5.
81. Casey, J.; Holden, N. The relationship between greenhouse gas emissions and the intensity of milk production in Ireland. J. Environ. Qual. 2005, 34, 429–436.
82. Lovett, D.; Shalloo, L.; Dillon, P.; O'Mara, F.P. A systems approach to quantify greenhouse gas fluxes from pastoral dairy production as affected by management regime. Agric. Syst. 2006, 88, 156–179.
83. Basset-Mens, C.; van der Werf, H.M. Scenario-based environmental assessment of farming systems: The case of pig production in France. Agric. Ecosyst. Environ. 2005, 105, 127–144.
84. Eshel, G.; Martin, P.A. Diet, energy, and global warming. Earth Interact. 2006, 10, 1–17.
85. Peters, G.; Wiedemann, S.G.; Rowley, H.V.; Tucker, R.W. Accounting for water use in Australian red meat production. Int. J. Life Cycle Assess. 2010, 15, 311–320.
86. Peters, G.M.; Rowley, H.V.; Wiedemann, S.; Tucker, R.; Short, M.D.; Schulz, M. Red Meat Production in Australia: Life Cycle Assessment and Comparison with Overseas Studies. Environ. Sci. Technol. 2010, 44, 1327–1332.
87. Williams, A. Determining the Environmental Burdens and Resource Use in the Production of Agricultural and Horticultural Commodities; Defra: Cranfield, UK, 2006.
88. Millward, D.J.; Garnett, T. Plenary Lecture 3: Food and the planet: Nutritional dilemmas of greenhouse gas emission reductions through reduced intakes of meat and dairy foods. Proc. Nutr. Soc. 2010, 69, 103–118.
89. Saarinen, M.; Kurppa, S.; Virtanen, Y.; Usva, K.; Mäkelä, J.; Nissinen, A. Life cycle assessment approach to the impact of home-made, ready-to-eat and school lunches on climate and eutrophication. J. Clean. Prod. 2012, 28, 177–186.
90. Horrigan, L.; Lawrence, R.S.; Walker, P. How sustainable agriculture can address the environmental and human health harms of industrial agriculture. Environ. Health Perspect. 2002, 110, 445–456.
91. Pimentel, D.; Pimentel, M. The Future of American Agriculture. In Sustainable Food Systems; Knorr, D., Ed.; Avi Publishers: Westport, CT, USA, 1983; pp. 3–27.
92. Reijnders, L.; Soret, S. Quantification of the environmental impact of different dietary protein choices. Am. J. Clin. Nutr. 2003, 78, 664S–668S.
93. Hoekstra, A.Y.; Chapagain, A.K. Water footprints of nations: Water use by people as a function of their consumption pattern. Water Resour. Manag. 2007, 21, 35–48.
94. Morgan, E. Fruit and Vegetable Consumption and Waste in Australia; State Government of Victoria, Victorian Health Promotion Foundation: Victoria, Australia, 2009.
95. De Boer, J.; Schösler, H.; Boersema, J.J. Climate change and meat eating: An inconvenient couple? J. Environ. Psychol. 2013, 33, 1–8.

96. Pimentel, D.; Pimentel, M. Sustainability of meat-based and plant-based diets and the environment. Am. J. Clin. Nutr. 2003, 78, 660S–663S.

97. Baroni, L.; Cenci, L.; Tettamanti, M.; Berati, M. Evaluating the environmental impact of various dietary patterns combined with different food production systems. Eur. J. Clin. Nutr. 2007, 61, 279–286.

98. McMichael, A.J.; Powles, J.W.; Butler, C.D.; Uauy, R. Food, livestock production, energy, climate change, and health. Lancet 2007, 370, 1253–1263.

99. Ausubel, J.H.; WernIcK, I.K.; Waggoner, P.E. Peak farmland and the prospect for land sparing. Popul. Dev. Rev. 2013, 38 (Suppl. 1), 221–242.

100. Williams, J.E.; Price, R.J. Impacts of red meat production on biodiversity in Australia: A review and comparison with alternative protein production industries. Anim. Prod. Sci. 2010, 50, 723–747.

101. Chassot, A.; Philipp, A.; Gaillard, G. Oekobilanz-Vergleich von intensivem und extensivem Rindfleischproduktionsverfahren: Fallstudie anhand zweier Fallbeispiele. (in German). In Proceedings of Ende der Nische-Beiträge zur 8. Wissenschaftstagung Ökologischer Landbau, Kassel, German, 3 January–3 April 2005.

102. Johnson, J.M.-F.; Franzluebbers, A.J.; Weyers, S.L.; Reicosky, D.C. Agricultural opportunities to mitigate greenhouse gas emissions. Environ. Pollut. 2007, 150, 107–124.

103. Nguyen, T.; van der Werf, H.M.G.; Eugène, M.; Veysset, P.; Devun, J.; Chesneau, G.; Doreau, M. Effects of type of ration and allocation methods on the environmental impacts of beef-production systems. Livest. Sci. 2012, 145, 239–251.

104. WWF. The Growth of Soy: Impacts and Solutions; WWF Internationa: Gland, Switzerland, 2014.

105. Cassman, K.G. Ecological intensification of cereal production systems: Yield potential, soil quality, and precision agriculture. Proc. Natl. Acad. Sci. USA 1999, 96, 5952–5959.

106. Tilman, D.; Cassman, K.G.; Matson, P.A.; Naylor, R.; Polasky, S. Agricultural sustainability and intensive production practices. Nature 2002, 418, 671–677.

107. Fadel, J. Quantitative analyses of selected plant by-product feedstuffs, a global perspective. Anim. Feed Sci. Technol. 1999, 79, 255–268.

108. Nepstad, D.C.; Stickler, C.M.; Almeida, O.T. Globalization of the Amazon soy and beef industries: Opportunities for conservation. Conserv. Biol. 2006, 20, 1595–1603.

109. Fearnside, P.M. Soybean cultivation as a threat to the environment in Brazil. Environ. Conserv. 2001, 28, 23–38.

110. FAO. The State of World Fisheries and Aquaculture; FAO Fisheries and Aquaculture Department: Rome, Italy, 2012.

111. Villasante, S.; Rodríguez-González, D.; Antelo, M.; Rivero-Rodríguez, S.; de Santiago, J.A.; Macho, G. All Fish for China? Ambio 2013, 42, 923–936.

112. McClanahan, T.; Allison, E.H.; Cinner, J.E. Managing fisheries for human and food security. Fish Fish 2013.

113. Beveridge, M.; Thilsted, S.H.; Phillips, M.J.; Metian, M.; Troell, M.; Hall, S.J. Meeting the food and nutrition needs of the poor: The role of fish and the opportunities and challenges emerging from the rise of aquaculturea. J. Fish Biol. 2013, 83, 1067–1084.

114. Pauly, D.; Chua, T.-E. The overfishing of marine resources: Socioeconomic background in Southeast Asia. Ambio 1988, 17, 200–206.
115. Hilborn, R.; Branch, T.A.; Ernst, B.; Magnusson, A.; Minte-Vera, C.V.; Scheuerell, M.D.; Valero, J.L. State of the world's fisheries. Annu. Rev. Environ. Resour. 2003, 28, 359.
116. World Bank. Fish to 2030: Prospects for Fisheries and Aquaculture; Publisher: Washington DC, USA, 2013.
117. Merino, G.; Barange, M.; Blanchard, J.L.; Harle, J.; Holmes, R.; Allen, I.; Allison, E.H.; Badjeck, M.C.; Dulvy, N.K.; Holt, J.; Jennings, S.; Mullon, C.; Rodwell, L.D. Can marine fisheries and aquaculture meet fish demand from a growing human population in a changing climate? Glob. Environ. Chang. 2012, 22, 795–806.
118. Lea, E.; Crawford, D.; Worsley, A. Public views of the benefits and barriers to the consumption of a plant-based diet. Eur. J. Clin. Nutr. 2006, 60, 828–837.
119. Boeing, H.; Bechthold, A.; Bub, A.; Ellinger, S.; Haller, D.; Kroke, A.; Leschik-Bonnet, E.; Müller, M.J.; Oberritter, H.; Schulze, M. Critical review: Vegetables and fruit in the prevention of chronic diseases. Eur. J. Nutr. 2012, 51, 637–663.
120. Dauchet, L.; Amouyel, P.; Hercberg, S.; Dallongeville, J. Fruit and vegetable consumption and risk of coronary heart disease: A meta-analysis of cohort studies. J. Nutr. 2006, 136, 2588–2593.
121. Dauchet, L.; Amouyel, P.; Dallongeville, J. Fruit and vegetable consumption and risk of stroke a meta-analysis of cohort studies. Neurology 2005, 65, 1193–1197.
122. Drewnowski, A.; Popkin, B.M. The nutrition transition: New trends in the global diet. Nutr. Rev. 1997, 55, 31–43.
123. Hall, J.N.; Moore, S.; Harper, S.B.; Lynch, J.W. Global variability in fruit and vegetable consumption. Am. J. Prev. Med. 2009, 36, 402–409.
124. Roy, P.; Nei, D.; Orikasa, T.; Xu, Q.; Okadome, H.; Nakamura, N.; Shiina, T. A review of life cycle assessment (LCA) on some food products. J. Food Eng. 2009, 90, 1–10.
125. Renouf, M.A.; Fujita-Dimas, C. Application of LCA in Australia agriculture—A review. In Proceedings of the 8th Life Cycle Conference, Pathways to Greening Global Markets, Sydney, Australia, 16–18 July 2013; ALCAS: Gold Coast, Queensland, Australia.
126. Foster, C.; Green, K.; Bleda, M. Environmental Impacts of Food Production and Consumption. Available online: http://www.ifr.ac.uk/waste/Reports/DEFRA-Environmental%20Impacts%20of%20Food%20Production%20%20Consumption.pdf (accessed on 9 June 2014).
127. Rab, A.; Fisher, P.; O'Halloran, N. Preliminary Estimation of the Carbon Footprint of the Australian Vegetable Industry; Department of Primary Industries Victoria: Tatura, Australia, 2008.
128. Maraseni, T.N.; Cockfield, G.; Maroulis, J.; Chen, G. An assessment of greenhouse gas emissions from the Australian vegetables industry. J. Environ. Sci. Health Part B 2010, 45, 578–588.
129. Carlsson-Kanyama, A.; Ekström, M.P.; Shanahan, H. Food and life cycle energy inputs: Consequences of diet and ways to increase efficiency. Ecol. Econ. 2003, 44, 293–307.

130. Carlsson-Kanyama, A.; González, A.D. Potential contributions of food consumption patterns to climate change. Am. J. Clin. Nutr. 2009, 89, 1704S–1709S.
131. Leitzmann, C. Nutrition ecology: The contribution of vegetarian diets. Am. J. Clin. Nutr. 2003, 78, 657S–659S.
132. Baumann, A. Greenhouse gas emissions associated with different meat-free diets in Sweden. Uppsala University: Uppsala, Sweden, 2013.
133. Worrell, R.; Appleby, M.C. Stewardship of natural resources: Definition, ethical and practical aspects. J. Agric. Environ. Ethics 2000, 12, 263–277.
134. Buttriss, J. Food security through the lens of nutrition. Nutr. Bull. 2013, 38, 254–261.
135. Macdiarmid, J.I. Seasonality and dietary requirements: Will eating seasonal food contribute to health and environmental sustainability? Proc. Nutr. Soc. 2013.

Author Notes

CHAPTER 1

Conflict of Interest

There are no conflicts of interest to report. The authors are employees of Agriculture and Agri-Food Canada. The opinions expressed are the Authors' own.

CHAPTER 2

Acknowledgements

The research leading to these results has received funding from the European Union's Seventh Framework Programme (FP7-REGPOT-2012-2013-1) under grant agreement No. 316167 (Project Acronym: GREEN-Agri-Chains). Moreover, the present scientific paper was partially executed in the context of the project entitled "International Hellenic University (Operation – Development)", which is part of the Operational Programme "Education and Lifelong Learning" of the Ministry of Education, Lifelong Learning and Religious affairs and is funded by the European Commission (European Social Fund – ESF) and from national resources.

CHAPTER 3

Acknowledgments

The authors express their gratitude to the farmers who participated in the study and thank the anonymous reviewers for their helpful comments. They also gratefully acknowledge the contributions of Andy Boland, Joe Kirk, Tim Mackey and Joe O'Toole and thank the farm auditors who carried out the on-farm surveys.

CHAPTER 2

Competing Interests
The authors declare that they have no competing interests.

Author Contributions
TCE and KC designed and supervised all experiments. AKB conducted the experiments and participated in the analysis of the data. VU drafted the manuscript and participated in the analysis of the data. All authors participated in the interpretation of the data. All authors read and approved the final manuscript.

Acknowledgements
Salaries and research support was provided in part by the Ohio Agricultural Research and Development Center (OARDC) and the Hatch grant (Project No. OHO01222). We would like to thank International dairy ingredients, Inc. (Wapakoneta, Ohio, USA) for providing us with the milk dust powder used in this study. Our thanks also go to Dr. Fred Michel, Department of Food, Agricultural and Biological Engineering, OSU, for helping us with HPLC analyses.

CHAPTER 5

Acknowledgments
The authors appreciate the support and cooperation of the Innovation Center for U.S. Dairy by providing funding, participating the interpretation and providing opportunities to interact with major cheese-manufacturing facilities. The Innovation Center also supports open access availability to advance research studies. The authors also thank three ISO panel reviewers for their constructive comments.

CHAPTER 7

Conflict of Interest
The authors declare no conflict of interest.

CHAPTER 8

Acknowledgement

This case study was conducted for the partial fulfillment of Master of Organic Agriculture, Wageningen University, Wageningen, the Netherlands. The author acknowledges support, suggestion and cooperation from the staffs of Wageningen University.

CHAPTER 9

Acknowledgement

This work was supported by CNCSIS–UEFISCDI, project number PCCA2 62/2012.

CHAPTER 10

Acknowledgement

The Authors wish to thank Mr. D. Antolini and Ms. R. Villa for their support in the laboratory activities.

CHAPTER 12

Acknowledgement

The work described in this paper was supported by internal funding from the University of South Australia. We thank the two anonymous reviewers for their constructive criticism and suggested improvements.

Conflict of Interest

The authors declare no conflict of interest.

CHAPTER 13

Acknowledgement

The authors are grateful to Pieter Jan Brandsma and Henk Oostindie for the description and analysis of De Hoeve, to Peter Damary for the description and analysis of NaturaBeef, and to Burkhard Schaer, Claudia Strauch, and Karlheinz Knickel for the description and analysis of Tegut and Rhöngut. We want to thank the entire SUS-CHAIN consortium for their critical though supportive comments on the conceptual framework. The research on which this paper is based was financed by the European Commission (contract no. QLK5-CT-2002-01349). The views expressed in this article are those of the authors and do not necessarily reflect those of the European Commission. The authors also wish to express their gratitude for the helpful comments of the reviewers of a previous version of this paper.

Index

volatile matter, 222, 224, 231, 234

W

waste, xv, xvii, xix, xxii–xxiii, 4, 6,
9–11, 17, 22, 26–32, 34, 41, 45, 52–
53, 56, 58–61, 66, 95, 97, 99, 101,
103, 105, 107, 109, 111, 113–115,
118, 124, 133, 172, 211–212, 219,
221–225, 227–229, 231–235, 249,
253–259, 267, 276, 278–279
  agrifood waste, 45
  postconsumer waste, 133
  waste disposal, 26, 28, 133
  waste minimization, 32
  waste prevention, 27
  waste reduction, 9, 26, 28
  waste treatment, 28, 66, 222, 234,
  255, 257–258
wastewater, xix–xx, xxii, 17, 29, 53,
136, 146, 243, 249–250, 253, 255,
257–258

wastewater treatment (WWT), xix,
29, 136, 243, 249, 255, 257–258
water, xv–xvi, xix–xx, 4, 6, 9, 12–14,
17–19, 28–31, 38–39, 53–56, 61, 97,
100–102, 111–112, 118, 123, 127,
135, 139–140, 143, 146, 204, 211–
212, 225, 232, 237, 239, 249–250,
263, 268, 271, 277
  groundwater, 14, 19
  water conservation, xx, 31, 140,
  146
  water content, 225, 232
  water vapor, 100
whey, xviii, xx, 96–97, 103, 111,
114–121, 123–125, 127, 129–131,
133, 135–137, 139, 141–147, 149
wool, 155, 164, 166–167, 169–172

T - #0830 - 101024 - C320 - 229/152/14 - PB - 9781774637012 - Gloss Lamination